Rinat M. Nugayev

THE PLANCK-EINSTEIN BREAKTHROUGH:

RECONCILIATION OF THE PIVOTAL RESEARCH PROGRAMS OF CLASSICAL PHYSICS

MINKOWSKI
Institute Press

Rinat M. Nugayev
Volga Region State Academy
420138 Kazan
Republic of Tatarstan
Russian Federation

Cover: Planck's photo: https://commons.wikimedia.org/wiki/File:
Max_Planck_1933.jpg#/media/File:Max_Planck_1933.jpg
Einstein's photo: https://commons.wikimedia.org/wiki/File:Albert_
Einstein_photo_1920.jpg

ISBN: 978-1-989970-16-4 (softcover)
ISBN: 978-1-989970-17-1 (ebook)

Minkowski Institute Press
Montreal, Quebec, Canada
http://minkowskiinstitute.org/mip/

For information on all Minkowski Institute Press publications visit our
website at http://minkowskiinstitute.org/mip/books/

CONTENTS

INTRODUCTION: THE EPISTEMIC IDEAL MODELS AND THE HISTORY OF SCIENCE RECONSTRUCTIONS

It is commonly held that the American philosopher, historian, and sociologist of science Thomas Samuel Kuhn, along with his pre-eminent adepts, stood at the origins of the path-breaking *'Historical Turn'* in the steady philosophy of science realm. Therein the assiduous researchers increasingly commenced to realize that *'to comprehend what science is, one needs to understand how it advances'*. The exponential growth of history-of-science trailblazing rational reconstructions, obliged to this turn, overshadowed for the first time distinctive and subtle epistemological ideas that stood behind them. It engendered the enticing impression that the works of the thrilling epoch of *"die Sturm und Drang"* and, above all, the stupendous *"Structure of Scientific Revolutions"*, first published in 1961, constituted a *downright* generalization of history-of-science impelling data. Whereupon there transpired the illusion that to do away with logical empiricism, it was enough to just punctiliously 'wipe your eyes' and behold with an unbiased, fresh gaze at the history of science. And amazing epistemological discoveries will spill out of abundance.

Nevertheless, sober elucidation of what significant epistemological results were obtained supplanted precocious enthusiasm. This, in particular, was convincingly evidenced by the collection of Thomas Kuhn's latest articles *"After the Structure of Scientific Revolutions"*, already published after his premature death.

"Now I think *we overemphasized the empirical aspect of our enter-*

prise (an evolutionary epistemology need not be a naturalized one)"
(Kuhn 2000, p. 95; my italics).

In the epoch-making collection, one of Kuhn's ultimate goals was
not only to indicate the changes in his epistemological position that
gradually took place after 1961 but also to manifest their **philosophical premises**. Accordingly, the ardent appeal to the philosophy of
Immanuel Kant ('Ding an sich'), Hilary Putnam ('internal worlds'),
to Martin Heidegger's distinctive philosophy ('being-in-the-world') is
frequent and apparent. Yet the ardent attempt relentlessly elicited the
other side of the coin.

Ever since *"After the Structure of Scientific Revolutions"*, it becomes more and more clear that even the amplest, detailed, elaborated, and the sophisticated doctrine of modern history & philosophy of science constitutes such a hotchpotch of diverse philosophical /
epistemological / sociological / psychological / historical views, biases,
values, and opinions that it makes its objective assessment and active
application in current research extremely difficult.

Surely, this is not accidental and in no way is Thomas Kuhn's or
somebody else fault. The puzzling situation is a downward reflection
of the 'hard fact' that real, 'alive' scientific revolution is a jolly intricate and inextricable *mixture* of the most varied (and often rather
unexpected) factors, ranging from the social-psychological features of
scientific communities up to the odd social-cultural 'atmosphere' in
which the members of the communities have to ply. Perchance, it will
take generations of philosophers, sociologists, and historians of science
to salve the problems. In my humble part, I shall long to indicate a
promising direction of future research activity only.

Discreet descriptive methodology vividly embodies the hopeless-
ness of searching for an effective theory-choice resolution version. Its
advocates reckon that the methodologist's professional business is to
describe punctiliously the criteria *actually* employed by scientists. They
maintain that cultivating methods, norms and ideal cognitive models
is a hopeless task.

For instance, the lucid falsificationist account of scientific history
as a process of proposal and subsequent refutation of different theories
is unacceptable for a descriptivist not because it contradicts historical

data, but since it is a confounded *theoretical* description.   A historical juncture is pictured from the professional standpoint of a historian
or an epistemologist, but *not* from a *real participant* at the events.

In my view, the descriptivist sober arguments against the normative methodology (see Maurice Finocchiaro's thought-provoking writings, for instance) are based on setting the logical-theoretical descriptions in sublime philosophical minds against the logical-historical descriptions in the mundane minds of historical agents. The descriptivist arguments would be immaculately correct if these descriptions
were equally complementary, equivalent views just held by different
scholars on peculiar historical events. However, I perceive that these
types of descriptions refer to different levels of comprehension. The
descriptivist picture refers to an *empirical* level, and the normative
standpoint - to a *theoretical* one. These two levels are so *inextricably*
linked that the subtle and sophisticated theoretical description relentlessly 'shows itself' through the descriptive one. Vague and subtle, so
to speak, 'laws of development of science' exist only as the aspects,
the facets of individual actions of certain scientists.

Let me scrutinize the interrelation of theoretical and empirical inextricable levels of methodological studies. To give a sense of this interrelation, I would prefer to consider an oversimplified example from
statistical thermodynamics - just to catch the important hallmark that
is perchance common for models both in natural science and in humanities.

After Kant, it is a platitude that a simple description of a period
of scientific development can help to ascertain an *empirical* regularity
at best. To elevate it to the ranks of a genuine theoretical proposition,
i.e. to bestow on it the honorable status of *universality*, we are bound
to convert the empirical regularity to an expression stiffly connecting
certain terms of *theoretical language*. To be more exact, let me consider
a guileless example borrowed from Stepin (2005): the renowned Boyle'
law.

Behold the Boyle fine equation derived on the basis of some sedulous observations (on the metahistorical level this corresponds to a
certain apparent regularity ascertained for a certain amount of theory-
choice processes). We have examined N gases (sedulously have in-

quired into the transitions 'Ampére-Maxwell', 'Lorentz-Einstein', and 'Planck-Bohr'). However, no increase in the number N of gases considered (additional investigation of the 'Bohr -Schrodinger' transition) can ever elevate the discovered empirical regularity up to universal law. *No* finite number of observations can guarantee against the regularity being violated in some future observation or experiment. Moreover, such a violation is inevitable in the distinctive case of gases under high pressures. This case is exactly described by the complicated Van der Waals equation.

To furnish the fine theoretical law describing the evident connection between pressure and volume of a gas, the immaculate **ideal model** must first be punctiliously constructed. The set of abstract objects making up this model are ideally rigid and infinitesimally small particles, the 'material points' that untiringly collide with each other. These imaginary particles move according to the laws of classical mechanics, vigorously striking the walls of the experimental vessel. As a result, the pressure of the gas P is defined as the sum force of their blows per unit surface area. Speculative mathematical inquiry into the relations between these abstract objects makes it possible to *theoretically reproduce* the renowned law PV = const previously established only experimentally. Consequently, the inductively obtained expression becomes a true theoretical law, correctly describing the behavior of sufficiently rare gases. The decrepit empirical equation PV = const is now appeared to be turned into a sublime theoretical law. Whilst the form of the equation remains the same, pressure P and volume V epitomize the fine relations between abstract objects of a theoretical language now, rather than the blunt relations between real empirical gases and their vessels.

Thus, since the predictive force of inductive generalizations is probabilistic, the extrapolation of an empirical regularity is not enough to set it above the ranks of hypothetical supposition. The required transition is possible only when the connection between the quantities in the empirical expression is reproduced as a result of operations on the abstract ideal objects that make up *the ideal model* of the particular domain of reality. To get the genuine law describing the connection between volume V and pressure of a gas P, the ideal gas model should be constructed and the empirical regularity rederived from this ideal model.

This is in no way to defend the 'naive realist' stale position asserting that the theoretical law obtained is a more accurate' picture of reality'. On the contrary, as Nancy Cartwright (1983) shrewdly put it, Nature on her own may be lacking profound and uniform regularities. Most of the regularities we possess may represent only properties of the subtle ways in which we construct our theories (and experimental devices - see chapter I) to brood on things. As for ideal models, Cartwright retorts that in several branches of quantum mechanics one easily and cheerfully uses a whole bundle of models of the same phenomena. Incidentally, nobody reckons that a model of the throng constitutes the 'whole truth'. Moreover, they may be even mutually *inconsistent*. Ever since the models constitute primarily the intellectual tools that help us to understand phenomena and build the corresponding pieces of experimental technology. They enable one to intervene effectively into processes and to create new and hitherto unimagined phenomena (see Ian Hacking's fine 1985 book for a further discussion).

Furthermore, our common idea of approximation consists in that, to avoid a mess, we write down an equation that is true only approximately. But although some approximations are leading *away* from the truth, there are far more approximations *towards* the truth. Phenomenological laws, obtained as a result of approximating the fundamental laws to concrete situations, are *closer to the truth* than the fundamental ones. For obvious reasons, most of the fundamental laws describe extremely oversimplified and idealized situations that can have nothing to do with actual experimental facts (see chapter I for details). For instance, the four-potential of the electromagnetic field in Maxwellian electrodynamics $A^i = (\mathbf{A}, \phi)$ has no *direct* empirical meaning serving only as a powerful means for the mathematical description of electromagnetic phenomena.

However, **if the basic idealized models of a theory can be jolly far from the truth, what are they for?**

- Firstly, they supply one with a theoretical *vocabulary* to describe the empirical data and to comprehend quite novel phenomena, new empirical data. Moreover, most of the components of an ideal model are *idealizations*, i.e. the abstract objects that represent *caricatures* of real objects. The caricatures possess only part of the properties of real objects. For instance, a material point of Newtonian mechanics shares only one property with a tennis ball – 'to have a mass'. Yet we resolutely insert the material point into the ideal gas model and then apply the laws of classical mechanics to its components striving to demonstrate that the property 'to have a mass' is necessary and sufficient for a model to reproduce a certain realm of mechanical events.

- Secondly, especially in the most advanced natural sciences, they help to *predict* new phenomena and to construct experimental devices to check the predictions (see the chapter I for details).

- Thirdly, if the idealized model cannot proffer a proper explanation, it continues to form a "system of mathematical propositions, deduced from a small number of principles, which aim to represent as simply, as completely, and as exactly as possible a set of experimental laws" (Duhem [1906], 1954, p.19).

- And the last, but not the least, advantage of having an ideal model consists in that the model can *mark the domain of validity* of our solid knowledge more exactly. For instance, the basic premise of Boyle's fine ideal model consists in that we take the gas molecules as material points that can interact only during the collisions. The premise indicates that Boyle's model can describe the behavior of rarefied gases only.

Similarly, the recipe for salving the theory-change situation inductively obtained at the expense of examining the *already resolved* theory-change situations can only be taken as an empirical regularity at best. To be called *universal,* i.e. to attain the status of a theoretical proposition, it must be tentatively derived from the epistemic **ideal model of the theory-change process.** This model should *theoretically reproduce* the theory-change process by analyzing the subtle relations between the following subtle structures.

1. The *cognitive subject* S (scientific community in general or an influential group of researchers);

2. the necessary *internal* and *external* conditions of its creativity;

3. the *results* of its cognitive activities O – vague ideas, brand new hypotheses, powerful and solid theories, etc.

The relations between the theoretical objects of the epistemic model should be described by the abstract philosophical principles commonly borrowed from multifarious epistemological doctrines. On the other hand, each ideal model is intended to treat *its own facet* of the inextricable mellow theory-change process, consistently and persistently ignoring those peculiarities of the scientific revolution that it regards as 'inessential'.

The specific nature of the epistemic model is expected to be determined by its system of basic abstract objects, which in turn are grounded on those empirically fixed features of the cognition process that necessarily lead to the theory-change. The fundamental ideal theory-change model can be as far from the truth as Newton's fundamental theoretical scheme (describing the relations between such exotic ideal objects as the 'material points', 'forces', and 'inertial reference systems') is. But a *partial theoretical model* constructed from the basic one to scrutinize the concrete theory-change situation may be closer to the truth, to 'what happened' in the history of science. By punctiliously taking into account the peculiarities of the theory-change process under consideration, one can construct a sequence of such ideal models that are closer *to the empirical level,* to what had been *perceived* by real participants at the events.

Furthermore, to reproduce the inextricable interaction of external and internal factors in a scientific revolution theoretically, one has to investigate first the workings of each factor *independently* of the others, in so to say 'purified form'. And only after that, he obtains an opportunity to construct a model tentatively describing the interaction of the social-cultural factors with the internal or 'cognitive' ones.

The description of the accomplished theory-change process, grounded on the ideal epistemic model, will inevitably constitute the history of events that were chosen and interpreted in a *normative* way. Such a description will represent a **reconstruction** of real history since it will embrace only those facts of scientific history that are important for the model. On being applied to the past, the ideal model will determine the *rational reconstruction* of the real theory-change process. Being applied to the present state of the development of science, it will set some problem-solving *standards*.

The budding analogy between the ideal models in natural sciences and humanities leads to the interpretation of theory-change models that can better be comprehended in terms of Max Weber's distinctive notions of *action, social action,* and historical and sociological *ideal types.*

The concept of *ideal type* is pivotal for Weber's methodology of historical and sociological cognition. For the celebrated author of *"Die Protestantische Ethik"* an Ideal Type cannot be derived from empirical reality but is sedulously constructed as a theoretical scheme. Weber shrewdly called the ideal types *utopias.*

"The more sharply and unambiguously the ideal types are constructed, the more they are aligned to this world, and the better they fulfill their task, - in terminological, classifying, and in heuristic respects as well" (Weber 1964, band 2, p.15).

Weber's ideal types are analogous to ideal models of natural science. Ideal types "may be as rarely met in reality as physical reactions that are calculated under the assumption of absolutely empty space" (Weber, 1964, band 2, p.10). Such substantial notions as 'economical exchange', 'homo oekonomikus', 'craft', 'capitalism', 'Christianity' are typically ideal constructs, serving as a means of describing real individual historical formations. Incidentally, one must make out the *sociological* and *historical* ideal types. Both have a "character of some utopia, emerging under conceptual amplification, isolation of certain elements of reality" (Weber 1964, band 2, p.190). But there are important differences though.

"Sociology, as has been presupposed many times, contrives the notions of types and looks for *general rules* of events, unlike history, which seeks to give a causal analysis... of *individual,* culturally significant actions, formations, persons" (Weber, 1964, p. 545, band 2; my italics).

In sociology, ideal types are more universal than in history and can be called 'pure ideal types'. For instance, a sociologist strives to construct ideal models of domination that can be found anywhere at any time. Sociological ideal types have the following hallmarks.

1. The human actions which they encompass are constructed as if they have a place in *ideal* conditions.

2. This ideal form is independent of local conditions of place and time. Hence, while the historical ideal type serves to isolate the causal relationships that occur at least *once*, the sociological ideal types isolate the relationships that exist *constantly*.

3. Sociology for Weber is a peculiar discipline that studies *social actions*.

"Action is the name given to human behavior regardless of whether it is an external or internal act, inaction or suffering, if, when, and in as much as those who act give it a *subjective* meaning. But 'social action' should be the name given to such an action which in meaning, - as implied by those who act, - is related to the behavior of others, which in turn maps out its course" (Weber, 1964, band 1,p.1; my italics).

'Social action' is a genuine 'atom' of Weberian wayward sociology and cannot be divided into pieces. Moreover, neither society nor other forms of collective organizations can be taken as authentic subjects of action. Only individuals can. Thus, Weber does not forbid the application in sociology of such vague and controversial notions as 'state', 'family', 'nation', etc., but he insists that one should not   forget that they are not genuine subjects of social actions.

Weber famously elucidates four types of action. The two of them are *irrational*, in the sense that they are not the results of deliberate choices. These are *traditional action*, guided by habit, and *affectual action*, proffered by emotions. The further two are types of rational action. The first is *wertrational action* that is undertaken as a result of belief in the ultimate value of acting in a certain way. The second is *zweckrational action* that is a calculation of the appropriate action to be taken to achieve the desired end. Weber contended that in *modern* society zweckrational action, precisely because it is calculable and predictive, becomes the dominating type of social action in all the pivotal realms.

This action is the most comprehensible one for us since it:

1. is aimed at achieving those goals that an individual is aware;

2. employs for this attainment the means that are considered quite adequate by the individual himself.

Weber does not take zweckrational action as universal or even of particular significance in empirical reality. On the contrary, it is an *ideal type* of action, a *paradigm* of social action, a scale for comparison with other types of action. Zweckrational action is a cognitive rather than ontological notion; it is a helpful *means of analyzing reality* but is not reality itself. The true behavior of an individual is commonly directed simultaneously towards *several* types of action; it contains them as its important aspects.

Let me now return to the theory-change epistemic model. Following Weber, this model constitutes the ideal sociological type of scientist's action. The sociological type stands in resolute opposition to the historical ideal types used by descriptivists. The epistemic model as a paragon of all the ideal types reconstructs the scientists' social actions as if they have a place in ideal conditions. Besides, the ideal norms it produces are independent of local conditions of place and time.

On the other hand, an ideal epistemic model serves as an important *scale* for measuring the deviations from ideal norms and as a useful *paradigm* for analyzing new theory-change situations taking place in science. More importantly, we must make up *two kinds* of ideal epistemic models - the Basic Ideal Model that is jolly abstract and is extremely far from reality and the Particular Ideal Models constructed from the Basic one for examining the concrete situations. (Just as Weber masterfully had extracted his ideal type 'Capitalism' from the concrete analysis given in his fascinating *'Die Protestantische Ethik'*). Note that an epistemic model describes merely an aspect of an *individual* scientist's activities and not the activities of the scientific community.

However, on introducing an ideal epistemic model that does not emerge directly from history-of-science data generalizations, one inevitably comes up against the following problem. Let us assume that we have two (or more) basic ideal models, i.e. two (or more) different ideal types of theory-change, which can be applied to a theory-change process. As I have already pointed out, to choose between two basic models, one has to compare them as the tools for analyzing concrete theory-choice situations. One must compare them as *instruments of comprehending*. Therefore, a definite theory-change period should be chosen, and two particular reconstructions of the period produced by diverse basic models should be advanced and compared. Yet which reconstruction should be chosen? Appealing to 'hard facts' can occur pointless since the facts, as a rule, are fabricated by the models themselves. Different models commonly elicit different facts. For instance, falsificationists will stress and multiply examples of theory refutation, such as Einstein's refutation of Newton's theory, Maxwell's refutation of the Weber-Neumann theory, etc. Inductivists will appeal to inductive generalizations such as Kepler's startling laws, Faraday's renowned laws of induction, and so on. Even when facts are shared, they are weighted differently from one model to another. Whereupon, how can the disparity between the models and the facts be detected when the facts are chosen by the models? In other words, what are the *meta-criteria* of choice?

Applying Max Weber's sober evaluation of scientific action as the closest to zweckrational action and his notion of the social ideal type, one can conclude that this particular epistemic ideal model should be chosen, according to which most history-of-science data can be understood from a *unified* stance, and most scientists' expert opinions about the given period appear to be correct. Moreover, the ability of the model not only to comprehend the established history-of-science facts but also to help to discover the new thrilling ones is also important.

In my view, many rational reconstructions proffered met with substantial difficulties in effectively resolving either the mellow theory-choice situation that took place in the past (e.g. the Lorentz-Einstein transition) or the current theory-choice situation (see Nugayev 1999 for details). I maintain that the reason for such deficiency lies in the fact that the epistemic models of theory-change processes, on which the reconstructions are based, are too ineffective instruments for comprehending the genuine process of scientific cognition.

Hence, in **chapter I** of the monograph a wayward yet comprehensible epistemic model aimed at an analysis of the sober reasons for mellow theory change in science is posited. According to the lucid model, radical breakthroughs in science sprung out not due to the odd invention of new-fangled paradigms or the unfathomable creation of novel ideas *ex nihilo*, but rather to the long-term groping and relentless processes of the reconciliation and interpenetration of 'old' research traditions preceding such breaks.

The remaining chapters are devoted to showing off the lucid model's heuristic asset in reconstructing the concrete periods of development of physics. In **chapter II** an origin of the so-called 'old quantum theory', represented by Planck's pioneering 1897-1906 papers, is unfolded. It is manifested, that the primary quantum-theoretical schemes were punctiliously constructed as sheer 'crossbreed' ones composed from ideal models and laws of Maxwellian electrodynamics, Newtonian mechanics, statistical mechanics, and classical thermodynamics.

In **chapter III** it is contended that Einstein's 1905 'annus mirabilis' constituted an acme of the second scientific revolution. To fathom in what felicitous ways Einstein's medley 1905 writings hang together one is bound to pay special tribute to his longed, renowned strive for unity evinced in incessant attempts to coordinate the profound research traditions of classical physics. Though Einstein's vehement efforts sprung out of Max Planck's pioneering attempts to comprehend electromagnetic phenomena through the lens of conceptual structures of thermodynamics and statistical mechanics. Einstein's foolhardy light quanta hypothesis and distinctive special theory of relativity turn out to be mere milestones of the unwinding of Maxwellian electrodynamics and statistical thermodynamics reconcilement research program.

The aim of **chapter IV** is to modify the received view on the general theory of relativity (GTR) genesis and advancement by taking into account common scientific practice of its functioning, the history of science data and philosophy of science arguments. The genesis of GTR as an instance of an epistemological model of mature theory change that hinges upon 'old' theories encounter and interaction is considered. The arguments are strengthened in favor of the tenet that the dynamic creation of GTR had been continually governed by internal tensions between two research traditions, that of special relativity and Newton's gravity. The encounter of the traditions, their interpenetration and intertwinement entailed construction of the hybrid domain at first with an irregular set of theoretical models. Step by step, on eliminating the contradictions between the models contrived, the hybrid set was put into order. It is contended that the main reason for the GTR victory over the rival programs of Abraham and Nordström was a synthetic character of Einstein's program. Einstein had put forward as a basic synthetic principle the principle of equivalence that radically differed from that of rival approaches by its open, flexible and regulative character.

In **chapter V** in the light of the lucid epistemic model, the process of reconciliation of theories of electromagnetic and weak interactions that took place in the second half of the XX century is scrutinized. It is manifested that the epistemological specificity of the episode in the history of science is primarily expressed in the fact that the unification was carried out as a part of the substantially synthetic program that aimed to put up a sequence of intricate theories of electroweak interactions as consecutive approximations to the Yang-Mills gauge theory. The main milestones of the implementation of the synthetic gauge program that culminated in the advancement of the so-called 'Glashow-Weinberg-Salam model' are examined.

Eventually, in **The Epilogue**, an example of GTR and QFT encounter is scrutinized. It is contended that, under the epistemic model, the real progress commenced with the hybrid model construction in the brand new crossbred realm called the 'Quantum Field Theory in Curved Spaces'.

14

1 A LUCID EPISTEMIC MELLOW THEORY-CHANGE MODEL

In physics as a practical science, it has frequently happened and still does happen that two theories, developed independently of one another, come into conflict when extended and must be mutually modified to remain compatible. In this mutual adjustment lies the germ of their further development into a complete unity, since the chief purpose of each science is, and always will be, the unifying of all its great theories into one that will embrace all the problems of that science and afford a solution to all of them.

Max Planck. The Relation between Physical Theories

Probably, as a kind of general supposition, it can be contended that those trends in the history of human consciousness turned out to be most fruitful where manifold ways of thinking had encountered. These ways of thinking are deeply grounded in diverse spheres of human culture, in different times, in the plural cultural milieu, or varied religious traditions. When they actually meet with each other, when they correspond to each other so that an intense interaction between them takes place, one hopes that novel and interesting discoveries will follow.

Werner Heisenberg. 'Physik und Philosophie'.

Abstract. A wayward yet comprehensible epistemic model aimed at an analysis of the sober reasons for mellow theory change in science is posited. According to the cognitive model, an origin of scientific revolution consists not in well-trodden discrepancies between bold theoretical speculations and stubborn 'hard facts', but deadly contractions of mellow theories from 'old' rival scientific research programs with each other. The latter inexorably lead to profound contradictions that *can* be most effectively eliminated in a more general (global) theory. The hallmarks of global theory, as well as the stages of its thorny construction, are scrutinized. The model is illustrated in a walk chiefly regarding intricate physics of the early XX-th century, the three 'old' mellow theories in the case being Maxwellian electrodynamics, Boltzmann's statistical mechanics and classical thermodynamics, and to general relativity genesis. Enticing modern examples, referring to electroweak theory contrivance and general relativity and quantum field theory craving reunion, are also involved. It is elicited that the philosophical ground of the developed epistemic model consists in 'internal realism' amended by the so-called 'coherent theory of truth'.

Keywords: mellow theory structure, mellow theory change, profound contradictions, Popper, Kuhn, Lakatos, Feyerabend, Putnam, Stepin, Planck, Einstein, Hawking, Unruh, Zel'dovich.

1.1 A Downright Introduction: What are the True Reasons for Mellow Theory Change in Science?

What are the *true* reasons for mellow theory change in science? According to the well-trodden standpoint, a theory-change process takes place because of a smashing 'critical experiment'. The experiment unfolds the blatant inadequacy of the 'old' mellow theory. The latter manifests itself either as a blunt contradiction to an unmerciful critical experiment or as a dubious trend to adapt to its results using obsequious auxiliary *ad hoc* hypotheses (Karl Popper).

However, the widely accepted 'critical experiment' model or conception is grounded, as a matter of fact, on the so-called '**standard hypothetical - deductive interpretation of a scientific theory**'. (The inextricable term was perchance coined by Frederick Suppe [1974], 1977). Those who had upheld this epoch-making standpoint (a part of the 'Received View') alleged that all the scientific theories are structurally similar to the interpreted calculi of axiomatic theories of mathematics. Whereupon, a scientific theory can be properly reconstructed so that all its diverse content may be logically deduced from a few initial premises. At this juncture, the refutation of any single result automatically represents the refutation of the initial hypotheses and, consequently, of the whole theory. One single critical experiment is sufficient to destroy the entire gorgeous theoretical edifice.

Yet times are changing, and today the majority of philosophers of science, after the untimely demise of logical empiricism, contend that the standard hypothetical-deductive interpretation of the scientific theory is blatantly *insufficient*. This interpretation cannot adequately grasp the large group of scientific theories, as these theories contradict accepted standard interpretation of canonical axiomatic reconstruction.

A so-called '**nonstandard**' interpretation has successfully replaced the standard one. The latter has yet to be fully defined, and its numerous partisans differ in their views significantly. However, almost all of them share the following strong tenet.

A scientific theory necessarily embraces *at* least three groups of hypotheses that differ from each other in their significant functions to be dwelled on below.

1. The *mathematical* hypotheses that constitute the mathematical apparatus of a scientific theory.

2. The *semantical* hypotheses that shape the deft models of a scientific theory.

3. The *fundamental* hypotheses that constitute the basic premises of a scientific theory.

Besides the third group of hypotheses, some additional, yet less grand and abstract, hypotheses are assumed during the long strip of theory construction. These indispensable additional hypotheses cannot generally be derived from the basic assumptions of the scientific theory. When masterly added to the group of initial premises, some of the additional hypotheses may furnish a particular theory, which has a direct relationship only to a relatively small amount of phenomena.

Whence it is no wonder that the adepts of nonstandard interpretation share the point of view that *no* basic theoretical assumptions can be *directly* compared with experimental/observational evidence. This comparison should be procured only via the hypotheses of the *intermediate* level.

Thus, the Standard Hypothetical-Deductive Model was exposed to be unsuitable, since it had imposed too over-strict constraints on the intricate structure of the scientific theory. Yet it possessed at least one indisputable advantage - the *simple clarity*. However, the forced adoption of the nonstandard interpretation unavoidably broaches the following subject. If no strict rules are governing the transition from basic theoretical premises to experimental results, **how can a mellow theory be refuted as a whole?** And if it cannot be refuted, then how - in the real history of science - did one mellow theory replace another? Were the scientists engaged in the turnabout process spurred by rational considerations? Were they right or wrong?

In the history of the philosophy of science, the holistic standpoint is commonly associated with the celebrated 'Duhem-Quine' tenet. As Willard V. Quine reliably substantiated,

"The *totality* of our so-called knowledge or beliefs, from the most casual matters of geography and history to the laws of atomic physics or even of pure mathematics and logic, is a man-made fabric which *impinges on experience only along the edges*. Alternatively, to change the figure, total science is like a field of force whose boundary conditions are experience. A conflict with experience at the periphery occasions readjustments in the interior of the field. Truth values have to be redistributed over some of our statements" (Quine [1951], 2000, p.345; my italics; see also Duhem [1905], 1954, p.187).

It looks rather inspiring that the renowned 'Duhem-Quine' tenet unexpectedly finds substantial epistemological support in a rather diverse social-cultural milieu. Viz., according to one of the sophisticated nonstandard accounts (Stepin 2005), a *mellow* scientific theory constitutes a set of propositions that exhibits the relationships between abstract theoretical objects of at least two types – the *basic* objects and the *derivative* ones. The set of basic objects makes up the array of basic initial idealizations (the Fundamental Theoretical Scheme or FTS) with no direct reference to experimental data.

For instance, the FTS of classical (Newtonian) mechanics is represented by the basic ideal model embracing such apparent *idealizations* as 'the Material Point', 'the Force' and' the Inertial System of Reference'. The derivative theoretical objects are formed ('constructed') from the basic ones according to certain rules. The sets of derivative objects constitute *partial* theoretical schemes or PTSs. Any mellow scientific theory gingerly expands, to describe each novel experimental situation, due to transitions from FTS to PTSs. Each construction of PTS from the FTS represents an intricate problem that cannot be bluntly reduced to a strict and trenchant algorithm. Hence, it seems impossible, at first sight, to refute the FTS.

On the contrary, if the PTS contradicts reliable experimental results, another one can always replace it. Then what are the true reasons for mellow theory change in science? Elicitation of a simple divergence between theoretical predictions and experimental results refutes the partial theories only but not the mellow ones. When such anomalies are discovered, PTS can always be modified to coordinate the whole mellow theory with experience. If a single, arbitrary anomaly can always be eliminated, one can just as easily eliminate two, three, and four and so on – one *after* another. Whereupon how many anomalies are necessary to renounce a mellow theory?

Obviously, an *infinite* number of anomalies should be involved. While a mellow theory can be adapted to incorporate the results of any *finite* number of experiments, it cannot accommodate an infinite number of them. However, what is meant by the proposition that the theory 'contradicts an infinite number of experiments'? - The fact is that theory contradicts *all* the experiments of a certain type, i.e. not only the experiments that have been performed already but also those of the future. However, it then follows that one mellow theory contradicts another! Only genuine theoretical law can predict the results of future experiments, which have yet to be carried out. A theory is conjured up not only to explain the existing experimental data but also to predict the new ones.

Without claiming to be substantiated, even these downright considerations suggest the dictum that a mellow theory can be refuted only when it is impelled to explain a 'critical experiment' on a par with *another* different and contradictory mellow theory. The inextricable problem situation arises from the encounter of two mellow theories. The latter, in the inexorable process of their expansion, meet the experiment that belongs to the domains of validity of both theories. An apparent example is the renowned experiments related to blackbody radiation that demanded their theoretical reproduction and explanation of the joint application of dainty Maxwell's electrodynamics, daring Boltzmann's statistical mechanics and immaculate classical thermodynamics (see chapter II for colorful details).

Consequently, the account of wonted interaction of a single 'old' theory with experience is insufficient to comprehend (i.e. to 'reproduce epistemologically ') the intricate process of mellow theory change. Besides, the intense interaction of several 'old' mellow theories should be taken into account. If the experimental anomaly occurs because of the inevitable encounter of several incompatible mellow theories, it cannot be withdrawn by common deceptive modification methods. It's effective making away demands the proper resolution of the cross-contradiction. Most radically, this can be done by constructing a more general theory, containing the old mellow theories as to its partial and considerably modified ones. Just to quote the startling epigraph to this chapter (Planck [1908], 1960, p.4).

The **ultimate aim** of the present monograph is to elucidate and advance the above heuristic arguments, showing off an' ideal' epistemic model of mellow theory change, and to manifest the model's heuristical asset in the history of physics. (Recall the introduction: an epistemic model is 'ideal' in the same way as 'the ideal model of a gas' embracing a throng of material points and representing the first approximation to the real situation). The contrived model is illustrated chiefly concerning the 'Planck-Einstein breakthrough' in the early XX-th century, the three 'old' mellow theories, in this case, being Maxwellian electrodynamics, statistical mechanics, and classical thermodynamics.

Incidentally, I hope that the rational reconstruction of this particular episode in the history of science is fascinating in itself. It both elicits an origin of both the 'old quantum theory' and the special theory of relativity and explains why they were contrived almost simultaneously; in some important respects, the reconstruction supersedes the existing pre-eminent accounts (chapters II and III). Further to the belief that the *context of justification* cannot be comprehended outside the *context of discovery*, I maintain that the special theory of relativity was just a stage in the startling advancement of the 'old quantum theory'. Whereupon, the true reasons why it was accepted cannot be comprehended without taking into account Einstein's writings (and the startling works of his contemporaries) in the quantum realm (chapter III). Special relativity and the old quantum theory were invented within the same program of Boltzmann's statistical mechanics, classical thermodynamics, and Maxwellian electrodynamics reconciliation. Or, in candor language, quantum and relativistic revolutions were strikingly simultaneous, since they sprung out of a **common origin** - the stout clash between the three mellow theories of the second half of the XIX-th century that constituted the 'body' of classical physics.

The very implementation of indispensable *reductionist* and *synthetic* research programs is brought about by the clash of mellow theories, which they are designed to soothe and even take away. Within these programs (partially) empirically - equivalent scientific theories are settled. The theories from diverse research programs are empirically - equivalent since they are built up out of the same pile of bricks (see Nugayev 1999 for details) yet in according to diverse architectural projects.

For example, Lorentz's dainty electron theory was worked out within the reductionist program, while Einstein's special theory of relativity (STR) - within a sheer synthetic program of mechanics, electrodynamics, and thermodynamics reconciliation. Being taken independently, the STR did not explain any new experimental fact. For obvious reasons (the second STR postulate), the predictions of the Lorentz theory were identical to that of the STR, so that it would not be possible in any case to distinguish between the two rival theories on experimental grounds. It is no wonder that most of Einstein's contemporaries had vividly discussed the 'Lorentz-Einstein electron model', brooded on the 'principle of relativity of Lorentz and Einstein', and so forth. At the time of publication of Lorentz's dainty, second-order theory (1904) the only data available to test these theories were Kaufmann's dubious measurements of the masses of slowly moving electrons. However, they were initially taken by the scientific community as contradicting *both* STR and Lorentz's theory. It took a whole year for Einstein to retort Kaufmann's paper (see chapter III). In the end, the program capable of effectively taking away the cross-contradiction ultimately prevails. Having tentatively compared the heuristic potentials of the reductionist and the synthetic programs, I favor discreetly the latter group since it has the following peremptory objective advantages.

1. Firstly, synthetic programs should provide a larger enhanced *empirically progressive shift* of problems solved than the reductionist ones.

2. Secondly, only these programs can rationally explain the application of the so-called hybrid or *crossbred* objects, which spring

from the coincident theories.

If one inquiries into the multifarious structures of two celebrated theories, - quantum theory and general relativity, – he/she can conclude that their global theoretical schemes sprung out of the subtle amalgamation of the crossbred theoretical schemes. This is in no way to demand that each cross-contradiction should be made away (in the stale Hegelian way) by blunt choosing a synthetic program. I do not maintain that all the cross-contradictions that took their places in the history of science were cut down at the root with the help of synthetic programs. It means only that synthetic programs have some *objective* alluring properties, and that is all that can be ascertained on the intense competition of synthetic and reductionist programs in general. At the same time, this is not to disparage the lead of experiments in science. On the contrary, the epistemic model proposed seems to elaborate further on the current standpoint that both theorists and experimentalists have significant breaks in their respective traditions, but they are not typically *simultaneous* (Pickering 1985; Galison 1987). One should recollect Einstein's solid statement of fact in famous '*Auto-biographical Notes*' concerning the origins of the Quantum Revolution:

"Fundamental crisis set in, the seriousness of which was suddenly recognized due to Max Planck's investigations into heat radiation (1900). The history of this event is all the more remarkable because, at least in its first phase, *it was not in any way influenced by any surprising discoveries of an experimental nature*" (Einstein 1949a, p. 37; my italics).

Briefly, theory advancement must have, up to a point, a life of its own. The sedate development of two pivotal cultures within science does not mean that the two cannot many a time and oft speak to each other.

On the other hand, the issue under consideration can be approached somewhat differently. In my view, the philosophical ground of the developed epistemic model consists in '*internal realism*' amended by the so-called '*coherent theory of truth*' (see the last parts of this chapter). The gist of internal realism can be elicited by its comparison with its obsolete '*metaphysical*' ancestor. The common tenet for both relatives consists in the downright assertion that the world is made up of some mind-independent objects. However, while for a stout metaphysical realist there exists only *one* true and complete description of 'the way the world is', the internalist pliable philosophers contend with the discretion that there is more than one 'true' theory or description of the world (Putnam 2001). 'Truth', in a subtle internalist approach, is an odd kind of *idealized* rational acceptability. It is some sort of *ideal coherence* of our beliefs with each other and with other experiences as those experiences are themselves represented in our belief system. Hence, truth is not blunt correspondence with a mind-independent or discourse-independent state of affairs. There is no 'God's Eye point of view' that one can stubbornly access. There are only diverse (and often antagonistic) standpoints of actual persons reflecting various (often hostile) interests, values, and purposes.

Consequently, internal realism does not peremptorily renounce the stale standpoint that there are experiential inputs to knowledge. Knowledge is not a solid bulk of information with no constraints except internal coherence. Yet internal realism does deny that there are inputs, which are not themselves to some extent shaped by our vague concepts, by the intricate vocabulary we seek to report and describe them, or any inputs, which admit of only one single description, independent of all conceptual choices. What makes a statement (or a set of statements) rationally acceptable is its coherence and fit: coherence of 'theoretical' (or less experiential) beliefs with one another and with more experiential beliefs.

Incidentally, the fact that the epistemic model seeks to elicit the profound breaks in theoretical traditions diminishes its unquestionable domain of validity in the history of science significantly. The perennial split between theory and experiment has become a vital part of the most highly cultivated part of science - of physics - only in the second half of the XIX-th century. Previously physics (save Ptolemy and Copernicus's mathematical astronomies) was an experimental science *par excellence*. Only after the Second World War were theoretical physicists as numerous as experimentalists were (Galison 1987). Hence, the developed epistemic theory-change model is not omnipresent and universal, as I would like. It is not directly applicable to, say, XVII-th century physics, but commences to function properly only with the establishment of the modern one.

I am aware that the developed model highlights only the logical-methodological facet of an immensely complicated process that cannot be described in general lacking psychological and sociological terms too. That is why I aim to open wider Pandora's Box only, yet not to slam shut it. However, the aspect investigated renders the following grand picture. The history of science can be looked through a fluctuating pattern of diverse - theoretical as well as experimental - research traditions. Of this entire medley, only those can survive that manage to match and even support each other.

For instance, the motley world of 'old', pre-Einsteinian XIX-th century physics was conceptually and socially fragmented. It was split on at least three pivotal research traditions belonging to Maxwellian electrodynamics, classical thermodynamics, and Newtonian mechanics. Traditions organized around diverse groups of phenomena engendered slight support for one another. The partisans of each distinctive theoretical tradition recognized with a gnashing of teeth the existence of the other but doggedly went their wayward ways. With the advent of quantum theory and relativity, conceptual reconciliation of worldviews was accompanied by a social reconciliation of practices.

Henceforth the guileless idea that runs through the whole monograph as a red thread is rather straightforward. The renowned simple and clear accounts that had sought to comprehend mellow theory change only on the *monistic* grounds (i.e. surveying the dismal advancement of a single theory, a single scientific research program, a single paradigm, etc.) had unfortunately failed. Hence, one is compelled to cling to a *'pluralistic'* epistemological background. To fathom the true reasons for mellow theory change one should scrutinize the intricate interaction, interpenetration, and intertwinement of theories, research programs, paradigms, etc. Every paradigm consists not only of trenchant theoretical propositions but of vague values, whimsical metaphysical premises, and implicit research patterns as well. How the interaction of all these queer components of the witch's cauldron should be perceived and punctiliously depicted? – I cannot procure a general and immaculate answer. Yet I insist that an account of the subtle *dialectic* of the 'old' pivotal research traditions for comprehending mellow theory change is indispensable.

1.2 Part one. What Inevitably Makes a Mellow Theory the Established One.

The indispensable preliminary stage of the cross-contradiction occurrence consists in the independent, self-contained development of several mellow theories with domains of validity of their own. By definition, a theory is called *a mellow theory* if all its abstract objects are organized into diverse systems belonging to *three* distinct levels - the level of **basic** objects, the level of **derivative** objects, and the level of **empirical** ones (Stepin 2005).

To begin with, the fundamental laws of mellow theory describe the relationship between its *basic* abstract objects. The relationship between the derivative objects is highlighted by the consequences of the fundamental laws. For example, 'the electric field at a point **E**', 'the magnetic field at a point **H**', and 'the current density **J**' are the basic theoretical objects of Maxwellian electrodynamics. The dainty Maxwell equations immaculately describe the relationship between them.

Newton's laws describe fine relations between the basic objects of Newtonian mechanics. 'The absolutely rigid body', 'the central field', 'the harmonic oscillator', etc. represent the derivative objects of Newtonian mechanics. Certain laws of Newtonian mechanics display the relationship between them: that is, by the laws of rigid rotation, movement in the central field, etc.

The basic objects constitute the solid *basis* of a mellow theory. This means that, for a derivative object to join the system of theoretical objects, it must first be punctiliously constructed from basic objects according to certain rules. Basic theoretical objects are of *independent* construction; that is, they cannot be constructed from each other.

The solid set of the mellow theory basic objects makes up the basis, i.e. the stable subsystem of theoretical objects. All the basic objects are apparent *idealizations* and cannot reside as real bodies (like tables and chairs). For example, the material point is trenchantly defined as a body free of dimensions. As for the other basic objects of Newtonian mechanics, it is discreetly assumed that the inertial system of reference can be isolated from external influence. Though in reality bodies that can be completely isolated do not exist at all (even the black holes). However, the basic theoretical objects can be compared with certain evident fragments of Nature. The material points may be compared with real bodies whose dimensions can be ignored in the problem-solving process (see, for instance, Landau and Lifshitz 1973). The force can be compared with the actual interactions that change the bodies' states of motion.

The derivative subsystems are *subordinated* to the basic one, but they are independent of each other referring to different fragments of the same domain of validity. Each subsystem is characterized by its own set of notions and mathematical equations that constitutes a special part (section) of the mellow theory. For instance, Newtonian mechanics consists of the following independent sections: 'the small-oscillations mechanics', 'the mechanics of rigid body rotations', 'the mechanics of movement in a central field', and so on. Each of these proverbial sections is characterized by its own subsystem of derivative objects. Each subsystem is a solid model of a particular type of mechanical motion (the small oscillations model, the rigid rotations model, etc.). Particular laws of classical mechanics depict relations between the components of the subsystem.

In general, the relations between a basic subsystem and a subsystem of derivative ones can be highlighted in the following way. A derivative system is obtained from the basis by a process of *reduction*. It means that a mellow theory is developed not by formal logical, mathematical means only but also through fine gedankenexperiments with abstract theoretical objects. The reduction is put into effect by punctilious analysis of the nature of the empirically fixed domain of validity. This domain can be sifted and 'seen through' the lens of an ideal model, formed by correlations of basic objects. According to the peculiarities of each concrete experimental situation, various constraints may be imposed on the basic system. This enables one to

28

transform the basis into a subsystem of derivative objects. Whereupon, the fundamental equations are then applied to the derivative
subsystems to become the partial laws. The informal nature of such
metamorphoses converts the inference into a special problem-solving
operation. The solutions to such problems are prudently included in
theory at its origin. To the theoretician bothered by applying theory,
they serve as helpful *patterns* for subsequent puzzle-solving activity.
Each problem is scrutinized in accordance with primary **paradigms**
(in Thomas S. Kuhn's sense).

In classical mechanics, the paradigm examples are represented by
the 'derivations' from Newton's laws: the small-oscillations law, the
movement in a central field law, the rigid body rotations law, etc. In
classical electrodynamics, the paradigm examples are sedulous deduc-
tions of the laws of Biot & Savart, Coulomb, Ampére, Faraday et al.
from Maxwell's equations.

Discreet construction of the derivative objects from the basic ones
enables one to compare theoretical knowledge with experience, to explain and predict the results of real experiments. To this end, an
empirical equation - an intermediate relation – is derived from the
partial law. In this equation, the special constructs are carefully intruded. In contrast to abstract objects, the newly born constructs are
no longer idealizations, and they can be compared with real bodies
now. These constructs are called the *'empirical objects'* and their systems – artful representations of empirical situations - are called the
'empirical schemes'. Empirical objects are not equivalent to real bodies. An empirical object represents not a concrete single body with
which an experimentalist operates but merely a *class* of such objects.
Consequently, an empirical scheme corresponds not to a concrete experimental situation but to a *type* of such situations. For example, the
empirical scheme of the Biot & Savare experiment with a magnetic
needle and a conductor refers to an experiment with any current in
the conductor and any small magnetic needle.

Surely, a mellow theory becomes the" **established** "one when the
necessary links between *all* three levels of its organization are properly
installed. It makes possible to apply a mellow theory as an effective
instrument for making definite predictions (see the corresponding part

of this chapter for important details). All the bonds between the three levels of an established mellow theory should be sufficiently *rigid*. The indispensable rigidity allows one to connect a prediction referring to the upper level with all the underlying levels of a mellow theory. Hence, it allows one to construct an experimental device to check the theoretical prediction. A new result, obtained in the course of the advancement of the mathematical apparatus, immediately influences all the levels of a mellow theory. Hence, a theory can predict, and the predictions can be verified. A mellow theory obtains the status of the *established* one when at least some of its predictions are exhibited to be successful.

1.3 Part two. Unpreventable Occurrence of the Cross-Contradiction.

The accomplishment of a mellow theory T_1 inevitably gives rise to questions about the relation of T_1's basis, viz., $\{\mathbf{B_1}\}$, to the system of basic objects $\{\mathbf{B_2}\}$ of another mellow theory T_2. Are the basic theoretical objects from both theories constructively independent? Alternatively, is it likely that $\{\mathbf{B_1}\}$ belongs to a subsystem of derivative objects of T_2 (or vice versa)?

Alas, the problem is as old as mountains. For instance, the alternative ontologies of Ptolemy (mathematical astronomy) and Aristotle (qualitative physics), all appearance, arose out of diverse practices. Ptolemy's wayward ontology was grounded in pagan religious meditative practices (the divine mathematics of the Skies), and Aristotle's solid worldview was entrenched in sensual and object-manipulative ones (the mundane physics of the Earth). For many centuries, the dualism had been 'frozen up' in the Aristotle-Ptolemy pagan cosmology until the Christian canon Nicolaus Copernicus blew the dubious cosmology up from within. The explosion was performed in a quite different (monotheistic) social-cultural context, and, as a result, Galileo set up a throng of corresponding research practices aimed at the efficient reconciliation   of mundane physics of the Earth with sublime mathematics of the Skies.

In general, it is impossible to answer the above questions without taking into account the following *constructional* peculiarities of the derivative objects.

1. The rules for building the derivative objects from the basis are not clearly formulated algorithms. They are vaguely determined by the problem-solving examples or 'paradigms' prudently included in a mellow theory at its origin (Ptolemy's '*Almagest*', Copernicus's '*De Revolutionibus*', Newton's '*Principia*', Maxwell's '*Treatise*', Einstein & Grossmann's '*Entwurf*', and so on).

2. Application of these rules for reducing the basis to a subsystem of derivative objects takes the peculiarities of empirical reality for granted. However, these peculiarities vary considerably from one realm to another.

3. When the physical theories are diverse, the construction rules differ substantially from each other being governed by different paradigms.

Points (1) to (3) exhibit how difficult it is to expose that T_1 is subordinated to T_2 or vice versa. Therefore, in common scientific practice, the simple conjunction of $\{B_1\}$ and $\{B_2\}$ is assumed to constitute a novel basis.

The true necessity of analyzing the interrelations between $\{B_1\}$ and $\{B_2\}$ emerges when both theories are needed to explain certain experimental data. It is discreetly assumed that the experimental data can be described by a system of derivative objects constructed from the basic objects of both theories. Such derivative objects will be called 'crossbred objects' or simply '***crossbreeds***'.

Note how Max Planck famously outlined the important lead of experiments for the construction of the crossbred object in a paper with the meaningful title "*The Relation between Physical Theories*" (Planck [1908], 1960, p. 83).

The crossbred objects are constructed from the T_1 basic objects. Whereupon, the crossbred system will be a subsystem of T_1 derivative objects. On the other hand, the crossbred objects are put up from the

T_2 basis also. Hence, the crossbred system will simultaneously belong to the crossbred systems of T_2. Consequently, the partial laws of both T_1 and T_2 will describe relations between the crossbred objects. Incidentally, several 'domains of reality' can exist, the description of which may necessitate the joint application of two mellow theories. (For the sake of brevity I shall only dwell on two theories; in reality, three or more theories are commonly applied, but all these cases without much labor can be reduced to two-theory ones). Hence, several crossbred systems may subsist simultaneously. The joint application of T_1 and T_2 for solving a problem will be called, applying in a sui generis way the Podgoretzky & Smorodinsky felicitous notion (1980), *'theories' cross'* while T_1 and T_2 will be referred as *'cross theories'*. The set of statements describing the relations between the crossbreeds will be called a *'crossbred theory'*.

For instance, the accomplishment of new-fangled Maxwellian electrodynamics gave rise to conundrums on the relationship between its basis and the system of basic objects of entrenched research tradition of Newtonian mechanics. The task of theoretically describing a black-body radiation spectrum, the electromagnetic radiation processes demanded the joint application of both theories (see Poincaré's farsighted lecture course "*Electricite et Optique*", Paris, 1890 and Max Planck's fine 1906 paper, p. 616).

Let me peer at the manifest examples mentioned above more thoroughly.

(a) For the theoretical reproduction of a blackbody radiation spectrum, J. Jeans (1905) had dwelled on the system of standing electromagnetic waves in a closed cavity. By taking these waves as a system of harmonic oscillators (i.e. by constructing the crossbred theoretical objects), he could apply a well-known law of statistical mechanics (the equipartition theorem). In this way, the dependence of blackbody radiation energy on temperature and frequency was elicited. The system of crossbred theoretical objects, the correlations of which constitute a model of black-body radiation, makes up a subsystem of classical electrodynamics (i.e. it represents the system of standing electromagnetic waves). On the other hand, the same model constitutes a subsystem of derivative objects of classical mechanics (an odd mechanical system with an infinite number of degrees of freedom).

(b) Lorentz's celebrated *Theory of Electrons*, which immaculately explained and predicted a vast number of phenomena referring to the 'electrodynamics of moving bodies', provides a fascinating example of the genuine crossbred theory. Initially, following the traditions of Maxwell and his followers (Lenard, Hertz), it was assumed that electric charges could be taken as a kind of perturbations of the ether. This enticing assumption was grounded on the key idea of Maxwell's electromagnetic theory that displacement current is identical to conduction current. Whence Lorentz was able to represent the density of electric current in the form of electromagnetic field flow through a cover. Nevertheless, owing to the influence of atomistic ideas, Lorentzian electrodynamics contained the notion of currents but taken as a system of electrons interacting with the electromagnetic field. The system of electrons, as a system of charged particles, is a subsystem of classical mechanics. However, as a throng of electromagnetic field sources, it is a subsystem of Maxwellian electrodynamics also.

(c) The distinctive hypothesis of the atomic structure that assumes the existence of the atomic nucleus was proffered by Nagaoka already in 1905 - before Lord Rutherford's renowned experiments with alpha particles. When contriving the planetary model of an atom, the positive charge was defined as an 'atomic nucleus', and electrons were taken as being 'on a stable orbit around the nucleus'. The system of crossbred theoretical objects, the correlations of which constitute the planetary model of an atom, is a subsystem of classical mechanics, and it is a vital component of the model of 'motion in the central field'. On the other hand, this system constitutes a subsystem of classical electrodynamics (the so-called Hertz's oscillator).

The statements from T_1 and T_2 describe relations between the crossbred objects. The crossbred objects belong to the subsystems of *both* theories. Hence, the operation of crossbred construction is identical to that of endowing derivative objects, belonging to both theories, with *novel properties*. These additional properties of the objects of the first mellow theory correspond to a brand new set of relationships transported from the second one.

Systems of derivative objects of each cross-theory were constructed before they met. Each of them is a peculiar summary of corresponding experimental studies carried out *independently* of investigations connected with another mellow theory. Therefore, it is not surprising that, because of the construction of crossbred objects, theoretical objects with mutually *incompatible* properties should transpire in the same subsystem of derivative objects of one of the crossbred theories. In the above case studies, the appearance of objects with mutually incompatible properties was characterized by physicists as a dreadful *'ultraviolet catastrophe'* (Paul Ehrenfest's apt notion), the exasperating *'paradox of unstable atom'* (Wilhelm Wien's notion), and so on. Let us dwell on those paradoxes in more detail.

(a) As a result of constructing the crossbred system, composing the inextricable model of blackbody radiation, the free electromagnetic field transpired to possess two diverse properties: 'to be a system of standing electromagnetic waves' and 'to be a mechanical system with infinite degrees of freedom'. Indeed, from the mechanical point of view, the electromagnetic field is a field of electric and magnetic forces that are continuously distributed in space. Such a weird system has infinite degrees of freedom in contrast to the finite number of degrees of freedom of a common body. The material point, for instance, has 3 degrees of freedom. Three coordinates determine its position while the electromagnetic field at a point is determined by all the electromagnetic forces throughout space, i.e. by an infinite number of magnitudes. Independently of Rayleigh and Jeans, and masterfully applying classical statistics, Einstein disclosed that, at an arbitrary but finite temperature, the density of the electromagnetic field energy should be infinite. This is quite natural since at an arbitrary finite temperature the same amount of energy (proportional to temperature) falls on each degree of freedom. However, an infinite electromagnetic field density is incompatible with the second principle of electrodynamics grounded on statistical mechanics. One can always extract energy from the cavity containing such radiation and set the 'perpetuum mobile' of the second kind to motion. Therefore, the property of thermal radiation 'to be a mechanical system with an infinite number of degrees of freedom' is apparently incompatible with its property 'to be a system of standing electromagnetic waves'.

(b) As was later punctuated by Albert Einstein,

"This is the basis [i.e. Maxwell's equations for electromagnetic field plus Newton's laws of motion for material points] on which H.A. Lorentz obtained his synthesis of Newtonian mechanics and Maxwellian field theory. The *weakness* of this theory lied in the fact that it tried to determine the phenomena by a *combination* of partial differential equations (Maxwell's field equations for empty space) and total differential equations (equations of motion of material points), which procedure was unnatural. The inadequacy of this point of view manifested itself in the necessity of assuming finite dimensions for the particles in order to prevent the electromagnetic field existing at their surface from becoming infinitely large. The theory failed, moreover, to give any explanation concerning the tremendous forces, which hold the electric charges on the individual particles. H.A. Lorentz accepted this weakness of his theory, which was well known to him, in order to explain the phenomena correctly at least in general outline "(Einstein [1936], 1960, p.306).

(c) The renowned planetary model of an atom was constructed as a system of crossbred objects. As a result, the electron was found to have a novel hallmark imported from classical mechanics: 'to move in a circular orbit around the nucleus'. However, like any accelerated charged particle, the electron must *continuously* radiate electromagnetic energy. The energy loss must unavoidably lead to the collapsing of the electron on the nucleus. Therefore, the property of electron 'to move in a stable orbit around the nucleus' is incompatible with the property 'to be a negative charge inside the atom'. This paradox, which had been first manifested in the Nagaoka model, was retained in the Rutherford model. Because of the paradox, the arrogant scientific community was rather skeptical about the Nagaoka dainty model but was forced to accept, with teeth-gritting, the phenomenological model of Lord Rutherford because of relentless experimental evidence.

A system of theoretical statements necessarily contains expressions regarding the relations between abstract theoretical objects. Therefore, the objects with mutually exclusive properties transpired in the system of derivative objects, should give rise to mutually contradictory statements in the cross-theories. Bearing in mind Podgoretzky and Smorodinsky apt notion (1980), I would like to denote the appearance of incompatible statements, when the theories cross, by *'cross-contradiction'*. The examples of such cross-contradictions of the physics of the XX-th century beginning refer to the blackbody radiation theory, electrodynamics of moving bodies, and to planetary atom theory.

(a) 'Heat equilibrium of radiation with matter exists' (a theorem following from the second law of thermodynamics; see Planck's renowned lectures on the theory of heat) and 'heat equilibrium of radiation with matter does not exist' (a consequence of Rayleigh-Jeans law, see Lorentz's 1909 celebrated book on the theory of electrons). A young participant at the first Solvay conference (1911) recalled,

"This argumentation, analogous to Jeans' analysis of thermal equilibrium, led, however, to the well-known *paradoxical* result according to which no thermal equilibrium is possible" (Niels Bohr).

(b) "What causes all these difficulties? Lorentz's theory contradicts the purely mechanical notions to which the physicists hoped to reduce all the phenomena of the Universe. Indeed, while there is no absolute motion of bodies in mechanics, and there exists a relative one only, in Lorentz's theory there is a peculiar state corresponding physically to an absolute rest state: the state when a body is immobile relative to the ether" (Einstein 1910).

(c) "The atom is stable" and "the atom is unstable".

1.4 Part three. An Effective Way of Cross-Contradiction Elimination.

The cross-contradiction results from the construction of the crossbred-objects. To eliminate the cross-contradiction in the most effective and reliable way, the theory T_3 must be created, deftly comprising both cross-theories to exclude the possibility of constructing crossbreeds from the basises of both theories. Theory T_3 will be called a *'global theory'*. According to the developed epistemic model, there are two ways of creating a global theory that bar each other: **reductionist** (R) and **synthetic** (S).

(R) Application of the *reductionist* method of creating a global theory is grounded on the natural assumption that the basises of both cross-theories refer to *different levels* of theoretical object organization. Hence, D_1, the domain of validity of T_1, makes up a part of D_2, the domain of validity of T_2. The basis of T_1 is superciliously called a '*true*' basis. T_1 itself is impudently alleged as a '*fundamental*' theory, while T_2 is humbly christened a '*phenomenological*' one. The authenticity of a phenomenological theory should be established by sedulous constructing its basic objects from the elevated basis of a fundamental theory, as well as by proving that its main laws obediently follow from those of a fundamental theory. Lastly, the basis of a phenomenological theory obsequiously takes its sordid place as a derivative system of the renowned fundamental theory. The possibility of constructing a phenomenological basis from the basis of a fundamental theory must also be substantiated. The relation between fundamental theory and a phenomenological one should be identical to that between the classical mechanics and one of its parts - the rigid body rotation theory, for instance. The intricate problems of constructing the phenomenological basis from the basis of a fundamental theory are of particular significance for a reductionist program. For obvious reasons, they are called '*the fundamental problems*' (Tisza 1963).

(S) The *synthetic* method of creating a global theory is grounded on the following strong assumption. The basic objects of both cross-theories are *constructively independent* of each other, though their basises belong to the same object-organization level. Hence, the exasperating cross-contradiction must be eliminated by creating a brand new system of *global* objects from which the basises of both cross-theories can be constructed. The fundamental laws of both cross-theories should be manifestly deduced from those of the global theory. Finally, the subordinate basises of T_1 and T_2 should take up their ultimate positions as derivative subsystems of the global theory.

What are the substantial differences between reductionist and synthetic ways of creating a global theory?

Firstly, the proper realization of the reductionist program rules out the application of crossbred objects. They are by the book, bell, and candle rejected by such a program. In contrast, the synthetic program resolutely rules out the using of the crossbreeds as derivative objects of the cross-theories only. This program farsightedly takes the crossbreeds as belonging to the *highest* level of theoretical systems organization. Moreover, the implementation of the synthetic program results in the creation of a brand *new* system of abstract theoretical objects.

On the contrary, the rules of reducing the basis to derivative sub-systems are strictly determined in the global reductionist theory by puzzle-solving examples contained in the fundamental theory. In contrast, there are no rules for constructing the basic objects of a synthetic global theory. Hence a synthetic global theory can be put up by amalgamating consecutive cross-basises according to a new *Weltanschauung* (a vivid example is the contrivance of Maxwellian electrodynamics – see Nugayev 2015).

Secondly, synthetic global theory can be constructed using the method of mathematical hypothesis - the thorny path of the electroweak theory is a suitable example. Viz., Yang and Mills (1954) had arrived at the farsighted idea that any field theory (and even the theory of strong interactions) can be modeled following quantum electrodynamics. In the trenchant language of group theory, gauge transformations that leave the Lagrangian of quantum electrodynamics invariant belong to U (1) unitary group. Yang and Mills sought to contrive a similar theory that would be invariant under local transformations of the group of strongly interacting isospin symmetries - a special unitary group SU (2). Members related to the isospin multiplets of hadrons performed the lead of the first two members of the Lagrangian, which in quantum electrodynamics related to electrons and protons. Now the wave function displayed a two-component field (p, n), representing a doublet of nuclear isospin. As in quantum electrodynamics, this part of the Lagrangian turned out to be non-invariant concerning gauge transformations. However, if one introduces spin 1 W-particles that form a triplet (W^-, W^0, W^+), then he can introduce such parts describing interactions of W-bosons with nucleons in the Lagrangian that its invariance regarding gauge transformations will be provided.

As a result, a jolly odd theory advanced by Yang and Mills began to be taken as a model (a 'paradigm') of a whole class of gauge theories, embracing even the theory of gravity (see chapter IV for details). Thus, in the electroweak case, the construction of a theory commences with a quest for fundamental equations. Only after they have been found can philosophical interpretation and empirical justification start. For this reason, some of XX-th century ways of building a theory may differ radically from those of the past.

Implementation of the reliable and resolute ways of eliminating cross-contradiction is grounded on two *equally reasonable*, but mutually *incompatible* presumptions. Therefore, these methods must be used in *alternative ideal types* of global theory construction programs - two reductionist and one synthetic. Since several crossbred systems can be contrived simultaneously, several synthetic subprograms can cohabit at the same time. (For instance, Bohr and Einstein's 'old quantum theory' sub-programs were carried out relatively independently, but subsequently merged into a single and powerful one). Each program creates its wayward sequence of scientific theories on the solid basis of one of the above natural assumptions. Each theory from the sequence is contrived by the addition of an auxiliary hypothesis to the preceding one. Each of the propositions considered above leads to the assumptions exclusive to the given program. Therefore, following Imre Lakatos (1970, p.119), I will call these fundamental assertions the **'hard cores'** of multifarious reductionist and synthetic research programs. The following statement makes up the hard core of synthetic scientific research program: 'the basic objects of all the cross-theories are constructively independent'. The opposite is true of the reductionist scientific research program (SRP). Besides, the reductionist 'hard core' indicates which basic objects constitute the 'true' basis. Neither a single crucial experiment nor a sequence of such experiments can certainly disclose which SRP - reductionist or synthetic - can eliminate the exasperating cross-contradiction efficaciously. Hence, each 'hard core' is *irrefutable*. For instance, a reductionist program is stigmatized as 'unable to eliminate the contradiction' only after the convincing demonstration of its incapacity to solve the fundamental problems. The vivid examples of genuine reductionist programs in the history of physics are indissolubly connected with James Maxwell and Hendrik Lorentz. Viz.,

"To be sure, Maxwell still tried to interpret his field theory mechanically using mechanical ether models. But these attempts receded gradually to the background following the representation of the theory – purges of any unnecessary trimmings – by Heinrich Hertz so that in this theory the field finally took the fundamental position which had occupied in Newton's mechanics by the material point" (Einstein [1936], 1960, p.305).

As for the Lorentz reductionist program, which was inverse to Maxwell's, the final is the following Einstein's dictum:

"Furthermore, there was one consideration which pointed beyond the frame of Lorentz's theory. In the environment of an electrically charged body, there is a magnetic field, which furnishes an (apparent) contribution to its inertia. Should it not be possible to explain the *total* inertia of the particles electromagnetically? This problem could be worked out satisfactorily only if the particles could be interpreted as regular solutions of the electromagnetic partial differential equations. The Maxwell equations in their original form do not, however, allow such a description of particles, because their corresponding solutions contain a singularity. Theoretical physicists have tried for a long time, therefore, to reach the goal by a modification of Maxwell's equations. These attempts have, however, not been crowned with success" (Einstein [1936], 1960, p.306).

Each subsequent, (n+1)-th version of a reductionist or synthetic sequence of theories represents a more adequate realization of a program than the anterior, n-th version. Each of these sequences verges on a certain limit or *ideal* of the global theory. It is this ideal that governs the direction of development of each SRP ideal type.

1.5   Part four. Can any Anomaly Set Forth an Intricate Problem Situation?

Can *any* anomaly set forth a real problem situation? – Apparently not. Otherwise, science could not proceed further rash explanations of two or three stale phenomena. What kind of anomalies contribute to the problem situations? What anomalies testify about the bla-

tant disagreement of several mellow theories with each other? - To
furnish a sober answer it is necessary to inquire into the link connecting the occurrence of the anomalies with the existence of the cross-
contradictions. It is necessary to unfold the subtle ways by which
the cross-contradiction influences the disagreement of theoretical statements with experimental data.

Let us recollect that, to explain the empirical data, a Partial Theoretical Scheme (i.e. a system of derivative abstract objects) is transformed into an **Empirical Scheme (ES)**. The ES constructs differ
substantially from the constructs of the Partial Theoretical Scheme
(PTS) since they are surely *not* idealizations now. They can be compared with real objects taking part in real experiments. (In what
follows in this part, I shall use the ideas of the relevant chapter of
Stepin's 2005 book).

For instance, let us dwell on the situation in Maxwellian electrodynamics with a magnetic needle which is near the conductor carrying
a current. To check the validity of the Biot-Savare law, one has to
compute the needle's angle of deflection. However, at first sight, the
Biot-Savare formula has nothing to do with real objects. It describes
the correlations of theoretical abstract objects only. The theoretical
objects involved are an 'infinitely small current' and a 'magnetic field
generated by this current'. Hence, the Biot-Savare equation cannot be
applied directly to explain empirical data. It is necessary to interpret
the corresponding quantities as relating to the *concrete* experimental
situation first. That is why a certain consequence of the Biot-Savare
law - the empirical formulae - is sedulously derived, and the novel
quantities are tentatively inserted into it. They characterize the needle's deflection angle and configuration of the integral conductor. The
empirical data should be compared with this formula only, yet not
with the genuine Biot - Savare law. Note that though the components of empirical schemes can be compared with real objects, they
are not *identical* to them. For instance, a real magnetic needle and
real conductor carrying a current possess many properties. Yet in the
Empirical Scheme, they are represented only by the following ones: 'to
be guided by the magnetic field', 'to conduct the electric current', and
'to have a definite configuration', respectively. Any ES component

can be compared not with a single real object or real experimental device, but with a *class* of such objects or experimental devices. For instance, the 'needle & conductor' simple empirical scheme refers to *any* experiment with a conductor and a magnetic needle.

An ES can be obtained from above and from below, i.e. not only as a consequence of a theoretical law but from statistical 'generalization' of empirical data as well. All the components interacting in real experiments - the object of the investigation itself and the corresponding experimental device - are shown off as having statistically invariant properties only. It means that they are laid out as possessing properties that manifest themselves only through many times and oft-produced observations. An ES displays the features typical for many experimental situations only.

Correspondingly, a subsystem of derivative theoretical objects can be compared merely with an ES class. Hence, any system of derivative objects constitutes an *invariant content* of the empirical schemes. The distinctive notion 'invariant content' should not be comprehended as referring to the notorious *inductive* method of derivation of theoretical laws. To obtain a genuine theoretical invariant, one should know beforehand that the set of empirical schemes certainly makes up a class. This class is easily visible only from above, only via discreet reduction of derivative subsystems to empirical schemes. Yet comparing the diverse empirical schemes with each other, one cannot conclude beyond doubt that they have common content. The latter is determined by the basis and the rules of derivative objects construction from the basis. The correlations of the derivative objects exhibit a specific idealized experiment reflecting the most general and salient features of experimental practice.

A system of crossbred objects belongs to the derivative subsystems of both cross-theories. Consequently, the occurrence of inconsistent statements in the crossbred theory is identical to the synchronous occurrence of inconsistent statements in both cross-theories. The latter is identical to endowing the derivative objects with novel properties, incompatible with the old ones. It was already punctuated that a system of derivative objects is an invariant content of a certain class of empirical schemes. That is why the occurrence of the cross-contradiction should lead not only to the contradiction between the crossbred theory and a single experiment but to the contradiction with the *whole class* of experiments.

For instance, the occurrence of the cross-contradiction between mechanics and electrodynamics in the crossbred theory already considered - the planetary atom theory - is identical to endowing theoretical objects of electrodynamics with new properties incompatible with the old ones. This leads to the following. The crossbred theory contradicts the experiments not only with certain types of atoms - with alkaline metals, for example, - but to the experiments with *all* the elements of Mendeleev's motley table as well. According to the predictions of the planetary atom theory, *all* atoms should be unstable.

The occurrence of the cross-contradiction in the theory of blackbody radiation leads to the contradiction of the crossbred theory not only with the renowned experiments of Lummer & Pringsheim or Rubens & Kurlbaum. The infinity of the blackbody radiation energy density led to the queer situation, when

"*all* the bodies should continuously radiate the electromagnetic waves, and not only outside, but inside also, until *all* the energy of the bodies is ultimately transformed into electromagnetic field energy and the temperature falls to absolute zero" (Landau 1958,p.81; my italics).]

Note that the cross-contradiction manifests itself not by common deviations of theoretical predictions from experimental results. No theory ever managed to withdraw all its anomalies. When the problem situation is set forth by the cross-contradiction, such odd anomalies transpire that are *out* of the domain of usual deviations common for the theory under consideration. Namely, these 'abnormal' anomalies attract the attention of the scholars first and the period of ineradicable '**crisis**' commences.

The abnormal anomalies differ considerably from the common ones not only in their magnitude. It was not rare in the history of physics when the anomalies were deftly withdrawn by judicial *non ad hoc* modifications of the partial theories. Yet in the particular cases of abnormal ('screwy') anomalies, the inability of the cross-theories to end with such anomalies constitutes a vital factor for the occurrence of the problem situation. Why a common anomaly cannot be taken as a deadly rebuttal of a theory, forsooth? - It is because such an object can always be inserted into a derivative theoretical system that can change the relations between other theoretical objects. However, if the anomaly is not an ordinary one but is generated by the cross-contradiction, the situation changes gradually. At least two diverse mellow theories have to grapple now with an exasperating anomaly. In this intricate case, a reconcilement of one cross-theory with the observation data will inevitably lead to strong deviations from the same data of the other cross-theory ('the swing case'). As a result, the cross-theories can be brought to an agreement with the experimental results only after the cross-contradiction winding up, i.e. first after the successful accomplishment of the global theory creation.

The developed model brings some novel light on the stale dispute concerning the role of crucial experiments. According to the down-to-earth empiricist standpoint, the smashing single '*experimentum crucis*' is quite sufficient to refute a scientific theory. According to the elevated pantheoretic doctrine, critical experiments for mellow theories do not exist at all. The empiricist vulnerable conception is grounded on the fallacious tenet: there exists the fairy observation language completely independent of theoretical propositions. On the other hand, the sublime pantheoretic conception exaggerates the theory-laidenness of observations. Hence, it is no wonder that, as the partisans of extreme standpoints, the empiricists and the pan-theorists are right only partly. According to the developed epistemic model, there can be *any* number of critical experiments but only in the problem situation set forth by the cross-contradiction. An anomaly, generated by the cross-contradiction, constitutes a smashing' crucial experiment'. Though it is crucial not in the renowned 'falsificationist' sense. It cannot refute finally - it can limit the domain of validity only. If the experimental data are apparently out of the crossbred-theory predictions domain, this indicates that to eliminate the deviations one has to put up the global theory first. The common method of intruding some auxiliary modifications in any of the cross-theories would be useless.

It is clear now why certain experiments were regarded as 'crucial' only with hindsight, only some years *after* they were performed. It is unnecessary to ask what experiments - of Rubens & Kurlbaum or Lummer & Pringsheim - did refute the classical electromagnetic theory of radiation. In reality, both experiments were crucial, as well as many other ones that have been forgotten up to now.

Besides the already mentioned, a long list of real crucial experiments includes Young's famous experiment, abnormal Mercury perihelion precession observations, and so on. Yet the renowned Michelson-Morley experiment is not an indubitably crucial one. A typical 'bland crucial experiment' surely helped to choose between the partial and humble theories of Fresnel and Stokes vividly debating about partial or complete ether drag. Both mellow theories were judiciously proposed within the same Lorentz- Langevin -Wien program.

Let me inquire into the last example in more detail. The stout argument of the empiricists is Lord Kelvin's prophetic speech, ardently devoted to Two Clouds 'that obscure the beauty and clearness of the dynamical theory'. Yet a more thorough analysis elicits that the celebrated protagonist of the Classical Physics spoke not on ' hard facts' and ' ultimate refutations' but on two exasperating *paradoxes*. Kelvin punctiliously considered the experimental evidence when illustrating the paradoxes only.

For instance, the renowned First Paradox ('Cloud') – *'Relative Motion of Ether and Ponderable Bodies'* - was dealt with even by Fresnel and Young. It ineradicably involved the intricate question, 'how could the earth move through an elastic solid, such as essentially is the luminiferous ether?' (Kelvin 1901, p.1). The First Paradox consists in the exasperating question: if 'ether does occupy the same space as ponderable matter' and if 'that ether is displaced by ponderable bodies moving through space occupied by ether', 'how then could matter act on the ether?' (Kelvin 1901, p.3). The notorious Michelson & Morley experiment is envisaged only as making dubious one of the solution of the paradox (that of Fresnel), and the highly controversial 'ad hoc' contraction hypothesis of Lorentz & Fitzgerald is arrogantly called 'brilliant' (Kelvin 1901, p.6).

No wonder that the renowned Second Cloud is nothing but the famous Maxwell-Boltzmann doctrine regarding the partition of energy. It had been called 'the greatest difficulty yet encountered by the molecular theory' even by James Clerk Maxwell in 1873. It is not surprising that the textbook blackbody experiments of Lummer & Pringsheim and Rubens & Kurlbaum are not even mentioned.

The occurrence of the abnormal problem situation (or of stern 'crisis' in Kuhnian terms) leads to gradual changes in the activities of scientists involved. They strive to exchange the 'old' mellow theory for new ones. The scholars are inspired by the promising ideas that will later be reconstructed by the historians & philosophers of science as synthetic and reductionist 'hard cores' of the corresponding 'research programs'. As a rule, they are not aware that the sequences of theories they propose belong to strict and definite Procrustean forms of reductionist or synthetic programs. Moreover, actually, they are often unaware that they are putting the global theory up.

1.6 Part five. How a Global Theory is Selflessly Put Up.

Is it possible to contrive the omnipotent global theory? Is the probability that scientists can stumble against the global theory in their chaotic wanderings significantly more than zero? Can the close-knit scientific community be compared to a flock of monkeys vividly inserting the whimsical texts on their antediluvian laptops? Is the probability to arrive at the global theory equal to the probability that the monkeys will print '*The Paradise Lost*'? Or the '*Tractatus Logico-Philosophicus*'?

Surely, a synthetic global theory is groped in reality by trial and error. Quite contrary to a reductionist global theory (with stiff and definite rules of construction), a synthetic mellow theory is contrived gradually. In the course of the construction, intermediate hypotheses are discreetly proposed that happen to possess apparent opportunities of providing constant empirically progressive problem shift in comparison with reductionist programs. Let me unfold them.

The basic system of theoretical objects and various systems of derivative objects as well possess two indissoluble assets (Stepin 2005). Firstly, a system of theoretical objects, a theoretical model is a sui generis *representation* (see Wartofsky 1979) of some class of objects under consideration. For instance, the solid basis of classical mechanics depicts mechanical motions with the help of the ideal model. The model represents the motions of real bodies by the motions of the 'material points' in the 'inertial systems of reference' under the influence of the 'forces'.

On the other hand, a system of theoretical objects makes up a model that seeks to portray the features of the real *experimental device*. This property permitted Vyacheslav Stepin to call the basic system 'The Fundamental Theoretical Scheme' (FTS) and the derivative system 'The Partial Theoretic Scheme' (PTS). For instance, the FTS of classical mechanics that represents mechanical motion by the motion of a material point in an inertial frame of reference under the influence of the forces describes a *gedankenexperiment*. The latter portrays the main hallmarks of multifarious mechanical experiments with real bodies. The basis of Newtonian mechanics is a wayward generalization of practical operations with real bodies moving along the inclined planes, operations with oscillating pendulums, with collisions of billiard balls, and so on. In particular, the small-oscillations PTS constitutes a model describing the various features of the experiments with oscillations of real pendulums, real stretched strings, etc.

Just imagine now that the reductionist program has been successfully realized. It means that the global theory was constructed in such a way that the place of the Global Basis was steadily occupied by the basis of one of the cross-theories. The basis of the second cross-theory became the subordinate system of its derivative objects. This means that the same system of theoretical objects turned out to imbibe the motley features of radically *diverse* measuring operations. Their difference is caused by the fact that the basic systems of both cross-theories had been created *independently* of each other and long *before* they met. They were contrived as distinctive generalizations of independent and quite different operations.

For example, successful implementation of the Lorentz-Langevine-Wien reductionist program (reduction of electrodynamics to mechanics) would portent the following. One of the derivative subsystems of Maxwellian electrodynamics, consisting of 'differentially small current' and 'magnetic field of a current', – simultaneously reflects the substantial features of the experiments with the bodies moving along the inclined planes, of the experiments with the rotations of rigid bodies, etc.

It is as plain as a pikestaff that the whimsical paradox springs out of the fact that the basis of one of the cross-theories is fatally unable to generalize the features of radically different measuring operations. The paradox can be made away, and it was - in Lorentz's fine research program, for instance, - by artful contriving the 'operationally indefinite' object. It helps to 'wash away' the operational foundations of the 'fundamental' theory to such an extent that it becomes able to accumulate the measuring operations of the 'phenomenological' theory also. The peculiar ideal object – the notorious 'ether' - was inserted into the reductionist program of Lorentz, Langevine, and Wien as a 'carrier of electromagnetic oscillations'. It is a vivid example of an operationally indefinite object. No physical experiment could determine the motion through the ether (see Einstein's 'Paradise' letter to Mileva cited in chapter III).

The whimsical construct was introduced into the system of abstract objects of Lorentz's theory surely not due to certain generalizations of measuring operations. It was inserted for the construction of a mechanical basis from that of electrodynamics. The material points (the 'particles') were to be regarded as 'ether perturbations', whereas the forces acting on them were to be tentatively determined via the tensions of the ether.

The price of such a subtle procedure may turn out to be rather high. It can lead to a radical decrease in the *predictive power* of a theory. What is the cause of this power, forsooth? If a mellow theory is *not* a direct generalization of empirical data, how can it predict the results of new experiments that are not even performed? How can a mellow theory forecast future?

According to Stepin (2005), this opportunity is grounded on the stiff connection between the theoretical basis and real experiments. The indispensable link of the basis with the experiments is intermediated by the systems of derivative objects. A basic object represents the characteristic features of the relations of the derivative objects that belong to the lower level of the organization of theoretical objects. Hence, to render a true operational definition of a basic object is to posit an *idealized* measurement operation, a gedankenexperiment, yet not the real one.

For example, the operational definition of electric field density **J** is provided not by the description of real experiments with electrometers and charged balls. It is procured by the sedulous description of the interrelations of the Maxwell theory abstract objects - 'electric field at a point' and 'test charge'. These abstract objects are the components of the partial theoretical schemes of Maxwell's theory. Besides, *their* operational status is determined now by real, and not by ideal, measuring operations. For instance, the 'test charge' is determined through such action of a massive charged body on the other, when the reciprocal influence of the second body on the first can be neglected.

The bond between the PTS and ES levels is installed because all the PTSs constitute the idealized schemes of real interactions observed in *real* experiments. All the bonds between all the levels of a mellow theory should be stiff ones. This rigidity allows one to connect a prediction referring to the upper level with all the underlying levels of a mellow theory. Consequently, the insertion of operationally indefinite abstract objects, common for reductionist theories, should lead to a considerable *decrease* of the predictive power. It does not mean, of course, that a reductionist program cannot predict at all. (A vivid example is the mechanical interpretation of heat).

Radical elimination of the cross-contradictions is chiefly possible in the theory that tentatively contains all the cross-basises in *different* systems of derivative objects. Only in synthetic global theory can the global theoretical scheme constitute a model that aptly combines all the salient features of idealized experiments with the abstract objects of *both* cross-theories. I had previously punctuated that the creation of the global synthetic theory is possible due to the constant empirically progressive problem shift.

It is possible to explain now why synthetic programs should provide a larger empirically progressive problem shift in comparison with reductionist ones. The reason consists in the successful utilization of the crossbred objects by the synthetic programs and their strict prohibition in the reductionist ones. Although the crossbred objects should be thrown with disgrace out of the cross-theories, they are bestowed with the honorable rank of global theoretical objects in synthetic programs instead.

Indeed, according to their definition, global objects are the same objects from which the basises of T_1 and T_2 are tentatively constructed. The global objects should necessarily contain information about $\{\mathbf{B_1}\}$ and $\{\mathbf{B_2}\}$, but the crossbred objects possess this hallmark too! They are constructed from $\{\mathbf{B_1}\}$ and $\{\mathbf{B_2}\}$. Moreover, since the construction of the crossbreeds from the basises is possible, the reverse process of reconstructing basises from the crossbreeds should be possible also. I have already underscored that in general it is possible to construct *several* systems of crossbred objects. Each crossbred system contains only *part* of the information about their basises. Only the whole set of various cross-systems possesses *all* the information about their random 'parents'. This set is always open for new elements since it is impossible to declare beforehand that all the cross-domains are known to us. A mellow theory can suffer unrestrained development by the construction of PTSs from the FTS for any new domain of validity. That is why the global system may occur only because of careful amalgamation of all the crossystems and cannot live a secluded life. It is always open for new crossbred systems, and the introduction of novel elements can endow the old ones with novel promising properties.

This epistemic scheme can be illustrated by the example of the transition from XIX- th century physics to the physics of the XX-th century. Viz., the modern quantum theory was created by the consequent generalization of Bohr's crossbred atom models, Einstein's crossbred semi-corpuscular radiation theory, STR, and so on. The process is still not finished up to the present.

Coordination of new-fangled STR with the celebrated Newton's theory of gravity ultimately led to the General Theory of Relativity (GTR) creation (see chapter IV and Nugayev 2018). Einstein's strenuous efforts to incorporate Newtonian gravity into the STR peculiar framework commenced already in 1907 when he had to prepare a detailed review for '*Jahrbuch der Radioaktivität*' (Einstein 1907; see Vizgin 1981 for details). His primary effort transpired to be surprisingly unsuccessful since simple-minded (though lorentz-invariant) generalizations failed to explain properly the anomalous perihelion motion of Mercury.

Nevertheless, let us return to the general epistemic scheme of global theory creation. Why *should* the dazzling synthetic theories empirically supersede the jogtrot reductionist ones? *The insertion of the crossbreeds engenders contradictions in all the cross-theories and forces them to accommodate each other, to 'interpenetrate' and to 'enrich' each other.* For instance, let us first turn to our secure example with the pivotal theories of XIX-th century physics. According to such a shrewd witness at the events as Max Planck,

"In fact, the impulse for an extension of mechanics had to *come from without*, and it came from the electro-dynamical theory" (Planck [1908], 1960, p.83; my italics).

After some pages, Planck amends the example, furnishing the following further jolly important additions:

"To-day [viz., circa 1908], it is generally accepted that an absolutely rigid mechanical theory of ether consistent with Maxwell's simple equations cannot be countenanced and this makes a *gap between classical mechanics and electro-dynamics that cannot be bridged*. It only remains to define exactly the range of applicability of the two theories *or to modify one of them*. The former soon proved to be impracticable, for mechanics and electro-dynamics overlapped on the question of motion of electrons and the manner of settling the controversies was shown for the first time by the discovery of deviations from the laws of classical mechanics, the deviations finding expression in the variable inert mass of the electron. *Einstein's theory of relativity contains a simple, complete solution of the problem of reconciling mechanics and electro-dynamics* in a quite general way, as far as it retains the practical essentials of the classical theory and still fulfills the demands of electro-dynamic calculations. The modification in mechanics due to the principle of relativity is the introduction of a new universal constant, entirely foreign to classical mechanics, namely, the velocity of light in an absolute vacuum" (Planck [1908], 1960,p.85; my italics).

However, not only STR turns out an unexpected result of the mellow theories' interpenetration. The other Planck's startling example is the old quantum theory itself (Planck [1908], 1960, p.85).

Secondly, the dynamic creation of GTR had been continually governed by internal tensions between two research traditions, that of STR and Newton's theory of gravitation. The encounter of the traditions, their profound interpenetration, and subtle intertwinement entailed coming into being the rival programs of the relativistic theory of gravity construction: of the programs of Abraham and Nordström – on the one hand – and Einstein's deft program – on the other. It should be stressed that the intricate relations between the GTR and its artful rivals were far more complicated in 1907-1915 than it may seem from the pestered 'truth-falsity' dilemma. Thus, one can contemplate with surprise the interlacement and *interpenetration* of rival 'incommensurable paradigms' into each other.

The relentless encounter rendered the construction of the hybrid domain at first with an irregular set of theoretical models. The rival approaches of Einstein, Abraham, and Nordström grew closer; and eventually, Einstein transformed them into complementary '*mathematical*' and '*physical*' strategies of a systematic research program that engendered the notorious 'Entwurf' and the GTR. Systematically, on tentative eliminating the contradictions between the models invented, the hybrid set was put into order by Einstein via the principle of equivalence. Likewise, as in the STR case, Einstein was able to freely juxtapose Nordström, Abraham's, and his non-metric theoretical schemes without reducing ones to the others.

Hence, it can be exhibited (see Chapter IV for details; see also Nugayev 2018) that the main reason for the GTR 'victory' over the rival programs of Abraham and Nordström was a synthetic asset of Einstein's program. This farsighted program did supersede the rival ones because it ingeniously *assimilated* some substantial ideas of the Nordström sober program (Nordström 1912, 1913a,b), as well as some felicitous premises of the judicious program of Abraham (1912a,b,c,d; 1915). Viz.,

1. Invention of the relativistic theory of gravity had dawned with the crossbred object construction in Einstein's 1907 farsighted paper. The crossbred object insertion – the discreet introduction of inertial and simultaneously gravitational mass ($E = mc^2$)

– led *perforce* to STR methods penetration into the Newtonian theory of gravity and to reverse penetration of Newtonian gravity methods into the STR. As a result, both theories were radically restructured and the corresponding changes in both of them were set up. The changes were epitomized in the particular sequences of crossbred models, the byproducts of the transformation performed.

(a) On the one hand, an inevitable consequence of the STR invasion into the Newtonian theory of gravity turned out to be Nordström and Abraham's discreet scientific research programs.

(b) On the other hand, no less relentless, owing to the Equivalence Principle, was the Newtonian theory invasion into the STR. It led to the sequence of Einstein's conspicuous papers on the generalization of the relativity principle and to spreading the principle not only on inertial systems of reference but on the various accelerated systems as well (Einstein 1912a, b; 1913a,b,c; 1914, 1915). However, the most valuable result of the hybrid theories of Nordström and Abraham consisted in that both theories maintained *very promising hints* on how the global theory could be put up. Hence, the climax of the stage was Einstein's daring proposal of the fecund Equivalence Principle that became one of the firm GTR heuristic foundations.

2. The basic GTR model was sedulously constructed due to the proper unification of the hybrid models of Einstein, Nordström and Abraham invented within diverse research programs. The asset of Einstein's style of reasoning was his open-mindedness; his renowned anti-dogmatism made it possible for Einstein to freely juxtapose Nordström, Abraham's, and his non-metric theoretical schemes without blunt reducing one to other.

3. It is this asset that helps to comprehend the true reasons for Einstein's victory over the rival programs of Nordström and Abraham. The 'Entwurf' and the GTR artfully 'incorporated' Nordstrom and Abraham's deft accounts in a higher-level theory. Thus, a true reason for Einstein's victory over the rival programs

does not consist in that he peremptorily rebutted the alternative accounts. On the contrary, it consists in that he subsumed them all into a new mellow and grand framework. Hence, the decisive drive in advancing the GTR came from the craving attempts to match physical and mathematical approaches, embodied in the Abraham, Nordström, and Einstein's crossbred theoretical models. A trustworthy explanation of Mercury's perihelion motion turned out to be a natural by-product of strenuous reconciliation efforts.

4. The monumental volumes (Renn et al. 2007a, b) are permeated with the idea that GTR "was not just a one-man affair and a lucky individual discovery but the result of a profound transformation of the extended system of knowledge of classical physics". According to the authors, the genesis of GTR can be taken as an example of a sheer 'Copernican process' that is repeatedly observable at decisive moments in the history of science.

"This process implies that marginal elements of a complex architecture of Knowledge torn asunder by inner tensions become the starting points for reconstruction as a result of which a new structure emerges that is, however, essentially composed of the building blocks already available. This process may be designated as a 'Copernican process'" (Renn and Sauer 2007, p.47).

In a sense, the reconstruction (Nugayev 2018) can be taken as an attempt to *explicate* and *furnish* the aforementioned doctrine of the 'Copernicus process', rendering a *mechanism* of simultaneous resolute transformation of 'old' classical paradigms. The hallmark of the mechanism is the construction of the crossbred systems from the basic theoretical objects of 'old' mellow theories. These crossbred systems are subsequently generalized to erect a structure of a novel mellow theory. A mellow physical theory can be empirically successful if it represents a kind of 'ein entwurf' (a draft) that can be explicated by the results of new experiments that increase its empirical content. In this way, GTR could become such a vehicle for the production of new gravitational propositions. It managed to reconcile and intertwine the 'old' classical traditions of Newtonian mechanics (including Newtonian gravity) and STR in such a subtle way that they ceased to contradict each other and their newly contrived fusion could successfully explain and predict new experimental data embracing relativistic astrophysics and cosmology.

Thirdly, in the modern GTR &QFT case, that will be examined in detail further, "the general theory of relativity and the quantum theory of fields, *taken together in a new synthesis*, have thus far *enriched* each other" (De Witt 1980, p.683; my italics). The enrichment manifested itself in the downright penetration of QFT into GTR (the gravitational field quantization). On the other hand, much less known reciprocal GTR penetration into QFT manifested in the discovery of thrilling 'Unruh *effect*' in flat space-time.

"By building a simple model of a particle detector, and accelerating it in the vacuum state in flat space-time, it is possible to show that such a detector behaves exactly as if it were immersed in a thermal sea of particles of temperature $T = 8\pi a/kc$" (Unruh 1980, p.154).

The budding encounter of T_1 and T_2 increases their domains of validity on D_1 and D_2 respectively. The domain of validity of T_1 expands owing to D_2, and vice versa. The domain of validity of both cross-theories becomes equal to two $(D_1 + D_2)$. For instance, the promising creation of foolhardy photon theory (and of startling special relativity) was due to intense interpenetrations of the following mellow theories.

1. Of statistical mechanics and thermodynamics (Willard Gibbs's and Albert Einstein's pioneering inquiries into the realm of statistical thermodynamics).

2. Of statistical mechanics and electrodynamics (Einstein's 1904 solid papers on fluctuations of the electromagnetic field).

3. Of thermodynamics and electrodynamics (Max Planck pioneering introduction of the notions of temperature and entropy of the blackbody radiation).

4. Of Newtonian mechanics and Maxwellian electrodynamics (STR basic postulates, corpuscular theory of light).

The amendment of Einstein's ideas by Louis De Broglie (see Pais 1980, for details) consisted in endowing the 'ponderous matter' with wave properties, etc.

Thus, the *crossbred objects constitute the queer 'channels' through which the ideas and methods of cross-theories interpenetrate.* Let us compare synthetic theories with reductionist ones, which strictly forbid the construction of the crossbreeds. Consider the most favorable for reduction case, viz., when the fundamental problems are successfully solved already, and the domain of validity of the global theory becomes larger since D_2 was successfully added to D_1. The 'fundamental' theory T_1 has penetrated the 'phenomenological' theory T_2, but the reverse process is strictly prohibited. That is why the domain of validity of the synthetic ideal is two times larger than the domain of validity of the reductionist one. Namely, that hallmark should provide the ultimate victory of synthetic programs over their reductionist rivals. Although each global theory should withdraw the cross-contradictions according to the model proposed, the *real* history of physics is more intricate. The process of global theory contrivance may last for centuries embracing new cross-domains with the cross-contradictions playing the lead of powerful 'driving forces' in the creation of new mellow theories.

For instance, in his celebrated " *'De Revolutionibus'*, a queer Catholic canon Nicolas Copernicus had audaciously pioneered (waywardly following Aristarchus of Samos) in taking the Earth as a common planet orbiting the Sun. Hence, he had contrived a crossbred theoretical model capable of descending the mathematical principles from divine phenomena to the mundane ones (further implemented by Galileo). On the other hand, through the same crossbred object, the principles of Earth physics were genuinely ascended (Newton) from mundane objects to the Skies (see Nugayev 2013 for details).

Likewise, in *A 'Treatise on Electricity and Magnetism'*, a pre-eminent Compendium, a whimsical Cambridge post-graduate James Clerk Maxwell had contrived a queer crossbred object – the displacement current – that enabled him to extrapolate the electromagnetic principles on the optical phenomena, and vice versa. His Copernican deeds consisted now in strengthening arguments for the identification of electromagnetic and luminiferous ethers and the invention of the *crossbred theory* with the displacement current capable of artful unifying the theories of electromagnetism and optics (Nugayev 2015).

In the case of a significant clash between GTR and QFT, the developed epistemic scheme can be animated by the following quotation.

"It is ironic that just at the time when the experimental tests are becoming possible, are being performed, and are verifying the predictions of the theory, the most determined assault on the structure of the theory itself is taking place. In the attempt to make general relativity compatible with the quantum theory of fields, many feel that the theory must be *altered* in some fashion. The most notable example of this is, of course, supergravity, with its addition of the gravitino" (Unruh 1980, p.153; my italics).

The most remarkable hallmark of the modern supergravity theories consists in that all of them were punctiliously constructed without experimental results. Yet

"the second reply to such criticisms is that we have already a powerful 'experimental' fact that needs to be explained, and that is the *logical inconsistency* of the two theories" (Unruh 1980, p.154; my italics).

When contriving the global theory, the experimental data are taken into account *indirectly*, being artfully embedded into pieces of theory. The lead of contradictions between Theory and Experiment is deftly played by the cross-contradictions between several theories.

However, the true 'hard fact' consists in that the contrivance of the Global Theory had already begun. Indecisive results including newfangled supergravity were obtained without *any* experimental results. This is not to constitute the complacent standpoint since too many steps are to be taken until one can allege that the subtle unification machine is already at hand. What is crucial in establishing the supergravity is its empirical justification. In my humble view, the latter should consist first in establishing the *operational status* of the concepts involved. This status can be achieved by ideal measurement procedures that are analogous to the pre-eminent Bohr-Rosenfeld procedures in quantum electrodynamics. To render an operational definition of a physical quantity, connected with a basic object, is to procure a description of idealized measurement procedures.

"It was only by *asking questions about measurements*, about how the various phenomena would *affect measuring instruments* such as real-model gravity wave detectors or observers falling through the 'singularity' at R=2M of the Schwarzchild metric, that these issues were resolved"(Unruh 1986, p.128; my italics).

It is no wonder that the most intriguing results were obtained namely in this direction of research. I have in mind first the startling Unruh effect:

"The lesson of general relativity was that the proper questions to ask of a theory are those that concern the behavior of detectors that have been fashioned to detect the effects that concern one. For example, in trying to decide whether particles are created, one needs to define a particle detector within the theory; i.e. some model of a physical system that will respond to the presence of a particle. In my attempt to apply *this philosophy* in the mid 1970s, I looked at the behavior of a simple model of a particle detector (1) for some quantum field, which, for simplicity, I took as a massless scalar field F" (Unruh 1986, p.128; my italics).

The operational analysis of the concepts involved is indispensable for making correct predictions.

1.7 Part six. The Philosophical Background of the Developed Epistemic Model: Internal Realism Amended by the Coherent Theory of Truth.

On the untimely demise of logical positivism, Scientific Realism has returned as a pivotal philosophical position. Yet many scholars, following Thomas Kuhn, Paul Feyerabend, Imre Lakatos, et al., strive to cope with the realist standpoint by historical analysis for insight into the structure and dynamics of scientific theories. They perceive that the traditional realist position should be somewhat amended to meet the requirements of the time. In which intricate directions should it happen?

To begin with, common realism necessarily involves the conjunction of two strong tenets.

(IT) **An independence thesis.** Our judgments answer for their truth to a world, which exists independently of our awareness of it.

(KT) **A knowledge thesis.** We can know which of these judgments are true.

According to the history of philosophy, Rene Descartes elaborated one of the stoutest classical counter-arguments against realism: if the external world is completely independent of our knowledge of it, how then trustworthy knowledge about it is possible?

An attempt to find the way out was made by the so-called '*Scientific Realism*' that had shown up within the realist epistemological research program with the 'hard core' molded by the (IT) – (KT) stout tenets. The crux of Scientific Realism consists in the premise that one should recklessly believe in the existence of the objects postulated by our most successful scientific theories like celebrated Newtonian mechanics, Maxwellian electrodynamics, Relativity, or Quantum Theory. Science is the highest authority to guarantee the existence not only of 'electron' and 'electromagnetic wave', but also of 'black hole', 'naked singularity', 'wormhole', 'M - brain', 'inflationary Universe', and so on.

In my view, the most convincing argument in defense of Scientific Realism is the '*no miracles*' one: realism 'is the only philosophy that doesn't make the success of science a miracle' (Hilary Putnam).

Regrettably, two strong counter-arguments against Scientific Realism were put forward.

(*a*) *The fatal underdetermination of theory by observational evidence.*

Suppose that two theories T_1 and T_2 (e.g. GTR and Abraham's deft nonmetric theory) are empirically equivalent, in the sense that their observational predictions are identical. Then no body of observational evidence (e.g. 'the bending of light' stale observation or the notorious 'anomalous Mercury perihelion shift') will be able to decide conclusively between T_1 and T_2. The well-known counter- argument, according to which one should always choose the theory that furnishes the better explanation, relentlessly plunges into sad historical obstacles. If one delves into the cases where scholars have embraced their 'best explanations' as true, these explanations have commonly turned out false. Recollect, for instance, Ptolemaic notorious astronomy, the dubious caloric theory of heat, the highly debatable ether theory in electrodynamics of moving bodies, and so on.

(b) *Pessimistic meta-induction from past falsity.*

This famous anti-realist argument is grounded on the mellow theory change in science ('scientific revolutions'). According to the meta-induction thesis, tentative reflection on the abandonment of theories in the history of science provokes the dismal expectation that our best current scientific theories will themselves be eventually abandoned.

In his 1996 masterful paper '*A Confutation of Convergent Realism*', Larry Laudan unfolded the basic premises of the so-called '*convergent realism*' that at the first sight promise to counter the pessimistic meta-induction tenet. Viz., successful (i.e. making substantially correct predictions) scientific theories are typically *approximately true*, and more recent theories are closer to the truth than the older ones (e.g. Newtonian mechanics and quantum one). Hence successive theories in any science will be such that they preserve the theoretical relations and the referents of earlier theories so that earlier theories will be limiting cases of later ones (classical mechanics and STR, GTR, and STR, and so on). Thus, acceptable new theories should perspicuously explain why their predecessors were successful as far as they were successful.

However, a more thorough analysis (partly grounded on Laudan's 1996 advances) enables one to take the convergent realism with a grain of salt.

Firstly, as Thomas S. Kuhn had manifested (Kuhn 1958, 1963, 1973),punctilious comparison of the grand *Weltanschauungen* of Aristotle, Newton, Einstein, Bohr,and Witten (the superstrings) enables one to arrive at the dismal conclusion that they are *not* like the snapshots of the same object made with the different orders of exposure. In what queer sense Aristotle's physics approximates Newton's one? Or vice versa?

Secondly, within an undulatory theory of light, the optical ether concept was an evident success in understanding reflection, refraction,interference, diffraction, and polarization, not to forget that the optical ether theories had successfully predicted some substantially novel phenomena. A famous example is A. Fresnel's striking prediction of a bright spot at the center of the shadow of a circular disc (Worrall 1989). It suffices to recall J.C. Maxwell who asserted that the ether concept was better confirmed 'than any other theoretical entity in natural philosophy'.

Thirdly, some sheer 'convergent realists' (Heinz Post, Noretta Koertge, et al.) may still profess that 'new' theories manage to retain the confirmed portions of the 'old' ones. Nevertheless, the Copernican astronomy did not retain the pivotal mechanisms of the Ptolemaic one like the notorious 'equant'. Newton's physics did not retain even most of the theoretical laws of Cartesian mechanics, astronomy, and optics. Einstein's theory of light quanta & STR did not retain the ether and the spherical wave radiation mechanisms connected with it. Moreover, the GTR was in some faint sense a return to the ether (see Albert Einstein's thrilling correspondence with Hendrik Lorentz).

Finally, to recap, the assertions of convergent realists that after the scientific revolution the laws of the 'old' theory miraculously transform into the limiting cases of the 'new' theory reflect their daydreams at best. Surely, it remains unclear *why* new theoretical ontology should smoothly transform into the old one when some artificial parameters change. For instance, under $(v/c) \rightarrow 0$ (v- velocity of the body, c – velocity of light) STR should convert to Newtonian mechanics. Or under $(S/h) \rightarrow \infty$ (S – action, h – Planck's constant) nonrelativistic quantum mechanics should convert to classical one.

Incidentally, the superposition principle in quantum mechanics famously does *not* transform into classical one under $(S/h) \rightarrow \infty$, and 'mass' does lose its ability to transform into 'energy' under $(v/c) \rightarrow 0$. Likewise, behold the odd transition from Einstein's GTR nonlinear field equations to the so-called 'weak-field approximation': it is insufficient to commonly leave the first members of the metric expansion. One is impelled to insert some substantial assumptions to transform the new ontology into the 'old' one (Zel'dovich and Novikov 1973). Undoubtedly, this does not exclude the particular cases when *some* important consequences of the new theory transform into the consequences of the old one, as is the stale case of classical and relativistic mechanics. For instance, the length of a moving body $l = l_0 \ \sqrt{1 - (v/c)^2}$ under $(v/c) \to 0$ transforms into l_0.

Paul Feyerabend shrewdly had put it (*"Against Method"*, 1975) that Niels Bohr himself, being the author of the 'correspondence principle' (1922), had warned that the 'asymptotic connection' between the classical physics and the quantum one,

"As it is interpreted in the correspondence principle, does not lead to the gradual elimination of the differences between the quantum understanding of the radiation phenomena and the ideas of classical electrodynamics; all that is asserted is merely an asymptotic reconciliation of numerical statistical results".

In my judgment, the Convergent Realism should not be resolutely withdrawn. On the contrary, it may be substantially modified at the expense of rejecting one or several assertions of the protecting belt of the realist epistemological research program and via the introduction of some brand new hypotheses. For discovering these helpful premises, one should turn to conspicuous alternatives of Convergent Realism.

Mercifully, John Worrall holds a different stand to 'save' realism that looks more farsighted. He shrewdly admits that even predictively successful past theories often contain significant errors. However, he asserts that this does not require a blanket rejection of *all* the scientific claims about the hidden mechanisms behind the observable predictions. He maintains that the history of science exhibits that past theories are characteristically wrong about the *nature* of the unobservable entities, but not about the *structure* of their behavior. In particular, he concedes that XIX-th century scientists were wrong to consider electromagnetic radiation as embodied in the 'ether', but quite right about dainty mathematical *equations* governing electromagnetism. As a result, one should believe in the *structure* of unobservable reality postulated by successful theories but carefully avoid any claims about the *nature* of that inextricable reality.

Fresnel, for example, completely misapprehended the inextricable 'nature' of light; nonetheless, it is no wonder that his theory enjoyed the empirical predictive success that it did. It is 'no miracle' because Fresnel's theory, as science later was convinced, attributed to light the right structure.

Unfortunately, there have been revealed some objections to Structural Realism (see Psillos and Papineau 1996 for convincing details).

The crux of their counter-argument consists in that restriction of belief to structural claims is no restriction at all.

The notion 'structure' is so vague that it can always be described via the 'entity' language (and vice versa); too often one cannot separate 'entities' and structures' by clear-cut, strict, and definite lines. Alternatively, to put in common words, 'structural realism is too metaphysically revisionary'.

I should add to counter-list the following remonstrance: the *mathematical structure is often lost in mellow theory change*.

An evident example is the Copernican Revolution. The latter consisted not in Ptolemy-Copernicus Weltanschauung metamorphoses only but in the construction of the whole set of heliocentric astronomy founded on the Newtonian mechanics. The Copernican research program's dazzling development was accompanied by gradual elimination not only of an 'equant' but also of the whole paraphernalia of epicycles and deferents that was crowned by Kepler and Newton's laws.

In my view, the above critical arguments cannot resolutely refute the hard core of the Structural Realism research program. They appeal for its modification, though. One of the possible ways out lies in the *nontrivial fusion* of the structural realism enticing program with other congenial approaches, first with the ones that flourished within the neo-Kantian XX-th century philosophical tradition.Its shibboleth consists in Poincaré's laconic tenet:

"The aim of science is not *things themselves*, as the dogmatists in their simplicity imagine, but the *relations between things*; outside those relations, there is no reality knowable".

Poincaré regarded the unobservable entities postulated by scientific theories as 'things in themselves' (Kant [1771], 1998). Yet he was a kind of 'revisionist', arguing that the latter can be known *indirectly* rather than not at all since it is possible to fathom the intricate relations into which they enter. It is jolly important, in the light of subsequent appeal to a coherent theory of truth, that in fact, Poincaré followed the Husserlian brand of neo-Kantianism. The latter was contrived to recover the *objective* or intersubjective world from the subjective world of private sense impressions:

"What we call *objective reality* is... what is *common to many think-ing beings* and could be common to all; the harmony of mathematical laws" (Poincaré 1906, p.14; my italics).

An enticing hallmark of structural realism is that it follows the Galilean trend of reasoning on physics. Galileo, in turn, not only fol-lowed Plato's "*Timeus*", asserting that the magnificent *Book of Nature is written in mathematical language.* In mortal combat with Aristotle, he sought to lead physics away from grasping the 'essences of things'. As the Florentine had put it, 'the search for essences, in my judgment, is a vain and hopeless kind of pursuit'.

Needless to say that the ultimate problem of the modern philosophy of science is '*to provide a theory of objectivity without a representational theory of truth*' (Ian Hacking).

The realist epistemological research program was contrived to ex-plain the success of science, yet it remains substantially ad hoc con-cerning that success. In my view, a promising way out is to *modify radically* the hard core of the program. In that respect, one has to appeal to Hilary Putnam's (2001) encouraging writings referring to the so-called '*internal realism*'. Putnam's startling inquiry into the problem of truth commences with outlining two grand philosophical perspectives in the history of Western epistemology.

(**MR**) **Metaphysical Realism.** The world consists of some fixed ineradicable throng of mind-independent objects. Nevertheless, there exists *only one* true and complete description of 'the way the world is'. Hilary Putnam (2001, p. 251) successfully calls this trend 'the externalist perspective, because its favorite point of view is God's Eye point of view'.

(**IR**) **Internal Realism.** The second perspective has no generally adopted name, whereupon Putnam refers to it as the *internalist* one. It is characteristic of this view to hold 'what objects does the world consist of? 'This is a question that it only makes sense to ask within a theory or description. The internalist philosophers enunciate that there is more than one 'true' description of the world. 'Truth', in this subtle approach, is a kind of *idealized rational acceptability*; it is some kind of ideal **coherence** of our beliefs with each other and with other experiences as those experiences are themselves represented in our belief system. Hence, truth is not the *correspondence* with a mind-independent (or discourse-independent) state of affairs. There is *no* God's Eye point of view that one can ever access. There are only diverse points of view of actual persons reflecting various interests, values, and purposes.

Internal Realism, which inextricably holds together subjective and objective aspects of truth, is grounded on Immanuel Kant's epistemology. The Reason is not an exact snapshot of Nature. On the opposite, the crux of Putnam's distinctive doctrine is relentless and continuous *interaction of Reason and Nature*.

Internalism does not reject the stale standpoint that there *are* experiential inputs to knowledge. Knowledge is certainly not a bulk of distinctive information with no constraints except internal coherence. Yet internalism does deny that there are inputs, which are not themselves to some extent shaped by our concepts, by the subtle vocabulary we apply to report and describe them, or any inputs, which admit of only one single description, independent of all conceptual choices. What makes a proposition (or a set of propositions) rationally acceptable is its *coherence* and *fit*. Coherence of 'theoretical' (or less experiential) beliefs with one another and with more experiential beliefs, not to forget coherence of experiential beliefs with the theoretical ones. Incidentally, truth is not equal to rational acceptability for the following reason. Truth is supposed to be a property of statement that cannot be lost, whereas justification can. Up to a point, truth makes up a certain *limit,* an *idealization* of rational acceptability. One speaks as if there were such contraptions as epistemically ideal conditions, and one calls a statement 'true' if it would be justified under such conditions.

The second basic hallmark of the whimsical truth concept of internal realism is its *coherent* character. The multitude of its adherents embraces not only such confounded neo-Hegelians as William Bradley, Henri Bosanquet, or Brand Blanchard but also such genuine analytical philosophers as Otto Neurath, Carl Hempel, Willard Quine, Donald Davidson, and Hilary Putnam himself. According to H.H. Joachim, truth is 'the systematic coherence which characterizes a significant whole' (quoted from Lynch 2001, p.99). Truth *is* the coherence, or *mutual support*, of our diverse beliefs.

Moreover (Walker 2001), the coherent theorist does not depart so far from common sense as to have to deny such stale tenets as 'true propositions correspond with the facts'. The partisans of the coherent theory of truth can okay such remarks quite freely. Nevertheless, they just will not consider them as expressing the *nature* of truth,nor will they take the 'hard facts' to belong to a metaphysically independent stout reality. On the contrary, on their view, the facts are themselves determined by the coherent system of beliefs.

Note that to maintain that the belief is true when it is a member of a coherent set is not to commit oneself to idealism. The various elements of the relevant system are not restricted to beliefs only but also embrace non-subjective aspects of lived reality, such as institutional practices, cultural traditions, and even the components of Husserl's 'Lebenswelt' concerning the events of our everyday life.

A curious objection to a coherent theory was famously put forward by Bertrand Russell ('Bishop Stubbs' argument). According to Russell, a coherence theory allows *any* proposition to be true since any proposition can be a member of some coherent system. Yet, according to Walker, Russell's objection misses the point, since coherence theorists maintain that truth consists in the coherence of some *specific* set of beliefs (see Pierce 2001, p.206 for details). Yet the most alluring hallmark of the coherence theory of truth consists in that it resolutely rejects the Cartesian stout distinctions between subject S and object O. Forsooth,

'If thought and things are conceived as related only externally, then knowledge is luck' (Blanschard 2001, p.104).

This principled desire to trespass the common boundary between object and subject, between the world, on the one hand, and human knowledge, experience, interpretive schemes, values, and the like, on the other hand, ascends to profound classical philosophical research tradition of Kant and Hegel. According to the latter, the most obvious problem with the binary picture is asserted to be that human beings can't remove all traces of our engaged concern with the world. Us mortals can't attain a God's eye elevated view. Hegel decidedly denounced this conclusion because he believed that the binary picture implicated in the peephole view is just as wrong in its metaphysics as in its epistemology.

"We are not peering through a peephole at the world but are always already in its midst. . . and knowledge is the outcome of sensuous human practices involving some aspect of this world within which we live for coherentism, knowledge is ultimately a product of phenomena that are immanent to human belief systems and practices, social organizations, and lived reality" (Alcoff 2001, p. 162).

Note that this anti-Cartesian research tradition was vehemently continued and strengthened in non-classical continental philosophy finding its *apex* in Martin Heidegger's celebrated 'philosophy of being' (Heidegger [1927], 1997; see also Heidegger 1971, 1972, 1978, 1995, 2001).

In my view, the most evident hallmark of the startling influence of Heideggerian epistemology on modern philosophy & history of science is its inextricable connection with Thomas Kuhn's conception of scientific revolutions (see Nugayev and Burganova, 2016).

1.8 Part seven. Summing-up.

After Heidegger and under his strong influence a substantial impact into the considered continental tradition made Michel Foucault who doggedly contended that truth isn't outside power, or lacking in power: 'truth is a thing of *this* world'. Society possesses a specific *'regime of truth'* of its own, its own *'general politics'* of truth: that is, the types of discourse which it accepts and makes function as true (Foucault 2001, p.318). According to Foucault, the history of mankind in general and the history of science in particular make up a series of 'discourses' or 'ideologies' that are resolutely changed due to *irrational* reasons.

In my view, this standpoint may be attractive for common history; yet for the history of science it seems rather dubious. Whereupon one should not conclude from Foucault's pioneering inquiry into the subtle dialectic of inner transformations of knowledge and power on micro-and-macro-levels that scientific truth and ideology are *identical*. Even Louis Althusser used to say, 'ideologies may be just or unjust, but not true or false'. On the contrary, the ultimate aim of science is to produce *Objective Knowledge*. The growth of objective knowledge in science indispensably consists in gradual and relentless 'purification' of knowledge (Planck) from dubious anthropic social-cultural, 'ideological' remnants.

Nevertheless, although Putnam rejects 'unbridled relativism', a coherent theory of truth necessarily contains a *modicum* of relativism. For instance, for the reason that the 'coherentness' of a view can be attained in many multifarious ways. Also, this hallmark is tightly connected with another impediment: where the alternative ontologies (on which the empirically equivalent theories are grounded) come from? According to the coherent account, they come into existence since one is engaged with and related to the external world in a *variety* of specific locations and with a variety of specific projects. However, all these should not lead one to 'absolute relativism'.

"Absolute relativism produces tolerance for other cultures, but it also removes the need to revise one's own beliefs and practices in light of their difference from others' (Alcoff 2001, p.175).

The actual situation one increasingly finds himself in modernity rarely allows for a noncommittal tolerance. In global juncture, modern societies are so interwoven that few if any exist separate from the others; we are increasingly interdependent yet mutually influencing.

Internal realism easily and cheerily salves the paradox of many empirically equivalent descriptions of reality by punctuating that they come into existence since we are engaged with and related to the external world in a *variety* of specific locations and with a throng of specific practices. (Recall, for example, the case of the Ptolemy-Aristotle research program in astronomy).

With a similar degree of easiness, internal realism solves the Second Paradox of historical meta-induction chiefly pointing out that the scientists belonging to changing paradigms 'live in different worlds' (T.S. Kuhn's renowned expression) that are *incommensurable* (the Kuhn – Feyerabend thesis). Beyond any doubt, the sequence of scientific internal grand 'Weltanschauungen' should not tend to a certain limit.

However, as it often happens, on solving one throng of problems one easily gets a pile of other ones. Viz., how scientific progress should be comprehended? Besides, what is behind the *objectivity of scientific knowledge*? Wherefrom should it take place within the internal realism and how can it grow?

In my humble opinion, to answer the questions one should turn to the epistemic model of theory change developed in this chapter. The model is grounded on the tenet according to which one never fathoms the mechanism of mellow theory change without grasping the subtle mechanisms of theory reconciliation that are tightly interconnected (see also Nugayev 1999). In certain respects, the epistemic model leads one out of the borders of internal realism. Any paradigm, research program, research tradition, etc. constitutes a 'local', internal worldview, – forsooth, - engendered by the peculiarities of the culture in which its adherent is immersed (Aristotle, Newton, Lorentz, Maxwell, Einstein, Bohr, et al.). An *encounter* of diverse paradigms leads to their actual interaction, interpenetration, and intertwinement. The contrivance of the crossbred theoretical objects, constructed from the basic objects of all the encountered theories, constitutes a *necessary link* of this interaction. The crossbred objects are important *channels* through which the interpenetration of the methods and notion of one paradigm into another takes place. As a result, the brand novel theories spring out that reconcile with one another quite better.

It should be stressed that the process of reconciliation leads to the gradual elimination of the anthropic, local components of scientific knowledge containing the remnants of contingent peculiarities of the genesis of the paradigm. (The most apparent example is the evolution of the 'force' concept beginning from Aristotle through Newton's abstract object and up to pseudovector in Einstein's relativity. Likewise, the evolution of the 'space' concept beginning from Aristotelian ether patch via Newton's infinite isotropic and homogeneous vacuum and Einstein's space-time up to Witten's 10-dimensional inextricable mathematical object.

In my view, the best illustration of this tenet is a thrilling patch of Maxwell's 1877 paper dedicated to Helmholtz. that begins as follows:

"Hence the ordinary growth of human knowledge is by accumulation round a number of distinct centers. The time, however, must sooner or later arrive when two or more departments of knowledge can no longer remain independent of each other, but must   be fused into a consistent whole. But though men of science may be profoundly convinced of the necessity of such a fusion, the operation itself is a most arduous one. For though the phenomena of nature are all consistent with each other, we have to deal not only with these, but with the hypotheses which have been invented to systematize them; and it by no means follows that because one set of observers have labored with all sincerity to reduce to order one group of phenomena, the hypotheses which they have formed will be consistent with those by which a second set of observers have explained a different set of phenomena. Each science may appear tolerably consistent within itself, but before they can be combined into one, each must be stripped of the *daubing of untempered mortar* by which its parts have been prematurely made to cohere. Hence the operation of fusing two sciences into one generally involves much criticism of established methods, and the explosion of many pieces of fancied knowledge which may have been long held in scientific reputation" (Maxwell [1877], 1952,p.592).

It should be added that the ideas of Maxwell's profound 1877 paper were generalized and elaborated by several scholars (see, for instance, Tisza 1963; Rovelli 2008; Nugayev 2013).

In the process of the interpenetration of the paradigms, different theories begin to coincide with each other more and more. Consider the blackbody problem, for instance (Nugayev 1996). For the case of the radiation in a cavity, the classical electrodynamics relentlessly led to the so-called 'ultraviolet catastrophe', while classical thermodynamics (and statistical physics after it) insisted on the accomplishment of the first law of thermodynamics (energy conservation law). The transformation of classical electrodynamics, when the electromagnetic field energy was taken in a statistical mechanical way, i.e. when the energy of Planck's oscillators was properly quantized, had accommodated – in this local field – the consequences of electrodynamics to the consequences of thermodynamics and statistical physics.

The next vivid example is the effective reconciliation by Newton of the mundane physics (Huygens's pendulum experiments) with Divine mathematics (Kepler's laws) that successfully described the motion of the planets around the Sun (Nugayev 2013). In the end, on revealing the cross-contradiction, a situation occurs when the consequences from different theories with different theoretical ontologies that describe the behavior of different crossbred objects begin to coincide. William Whewell, a polymath master of Trinity College, Cambridge, had shrewdly called it 'the *coincidence of inductions*'.

"That rules springing from remote and unconnected quarters should thus leap to the same point can only arise from that being a point where truth resides" (Whewell 1847, p. 65).

It is the growth (and even confluence) of the hybrid domains that constitute the progress of scientific knowledge, which shows itself up in grinding of the theories to each other, in reconciling and even unification of substantially diverse scientific theories. The crossbred objects and the relations between them form the condensations, the 'knots' (Cassirer) of scientific knowledge that are punctiliously conserved in mellow theory change processes. The most apparent examples are the heliocentric system, the planetary model of an atom, the vacuum model in QFT in curved spacetimes. On the contrary, such bizarre theoretical objects as Ptolemy's concentric spheres around the Earth or luminiferous ether of the electromagnetic theory of light, absolute time, or absolute space are lost in theory change since they *hamper* the unification process. They are too anthropocentric, forsooth.

The typical example of 'the daubing of untempered mortar elimination' principle for James Clerk Maxwell was

"the progress of science in Newton's time [which] consisted in getting rid of the celestial machinery with which generations of astronomers had encumbered the heavens, and thus 'sweeping cobwebs off the sky' " (Maxwell 1873a, p. 315).

Whereupon internal realism is inherently subjective up to a point since it sanctions arbitrariness in choosing the ways of breaking the world up into different objects. Yet because of an encounter of two or more subjective ways of breaking up the world, there appears such a novel way that is significantly more objective for the old ones. The demarcation line between the theoretical ontologies becomes more transparent and thin. For instance, the Aristotelian physics strictly separated natural and violent motions, as well as the terrestrial (sublunary) and celestial (eternal) worlds. A Newtonian theoretical ontology with its isotropic and homogeneous empty space was masterfully set up only after the resolute elimination of 55 spheres of the Ptolemy motley pagan cosmology conjured up to divide two Aristotelian worlds. Likewise, the sublime superspace of supersymmetry theories was established only after eliminating the demarcation line between the bosons and the fermions.

In our inextricable times, some new connotations have been affixed to unity and disunity. Instead of affiliation between unity and internationalism, liberal democracy, and a rational worldview, the axis unity/disunity has come to polarize around ideas of cultural autonomy in opposition to forces of homogenization, hierarchy, and domination.

The incommensurability claim (that Newtonian mechanics and Einsteinian mechanics would not meet) found expression in the abstract frameworks of Rudolf Carnap, in the distinctive paradigms of Thomas Kuhn, and the conflicting interest-theoretical accounts of the Strong Programme of Sociology of Science (Barry Barnes and David Bloor). Against this tenet Peter Galison (1987; see also Galison & Stump 1996) and Rinat Nugayev contend that the different paradigms, research programs, and research traditions work out 'local trading zones' (Galison) or 'crossbred domains' (Nugayev) in which they can successfully coordinate their practices. The stable totality of scientific knowledge, human culture, and human Dasein is provided not so much by a situation when a 'hundred flowers bloom' and a 'hundred schools of thought contend' simply co-exist. However, certainly not by a notorious situation when one theoretical tradition relentlessly suppresses the other ones (cultural imperialism). Perchance the true prosperity is reached when the co-existing traditions freely **interact** and gladly **interpenetrate** each other.

2 PLANCK'S 1899-1900 BREAKTHROUGH: THE COMMENCEMENT OF THE SECOND SCIENTIFIC REVOLUTION

However, the most curious thing is that the quantum idea should be born half a century ago when the kinetic theory of matter was created. This idea is intimately connected with the molecular structure of matter and is a specific reflection of it

Dmitry Goldhammer

In physics as a practical science, it has frequently happened and still does happen that two theories, developed independently of one another, come into conflict when extended and must be mutually modified to remain compatible. In this mutual adjustment lies the germ of their further development into complete unity, since the chief purpose of each science is, and always will be, the unifying of all its great theories into one that will embrace all the problems of that science and afford a solution to all of them. From this point of view, it can be said that the science, which is nearest its goal, is that one which combined the greatest number of its theories.

Max Planck

Abstract. An origin of the so-called 'old quantum theory', represented by Planck's startling 1897-1906 papers, is unfolded. It is elicited that the primary quantum-theoretical schemes were punctiliously constructed as sheer 'crossbreed' ones deftly composed from ideal models and laws of Maxwellian electrodynamics, Newtonian mechanics, statistical mechanics, and classical thermodynamics. In this pioneering amalgamation, Ludwig Boltzmann's seminal statistical ideas and subtle technique turned out to be crucial. Forsooth, tentatively deriving his celebrated blackbody radiation law, Max Planck had to take the experimental evidence into account. Yet it spurred him not 'to deduce from phenomena' punctiliously but to apply more theory instead. Moreover, the blackbody solid experiments incited Planck to apply the statistical technique for properly arranging the radiation data in increasing portions. Planck's hybrid and intricate theories epitomized the initial stages of an ambitious program of Maxwellian electrodynamics and statistical mechanics effective reconciliation.

Keywords: Max Planck, Boltzmann, old quantum theory, blackbody radiation, reconciliation, Maxwellian electrodynamics, statistical mechanics, classical thermodynamics.

2.1 Introduction. Controversy Around the Old Quantum Theory Genesis.

Many introductory textbooks on quantum mechanics and several corresponding history & philosophy accounts of the old quantum theory origins (see, for instance, Norton 1993) are grounded on explicit or implicit assumptions according to which this theory was contrived as a result of a downright 'deduction from phenomena'. Furthermore, the so-called 'old quantum theory' in the academic history of science has been taken as *paradigmatic* of a case where theory is completely determined by the evidence. According to the pre-eminent narratives, the 'crucial experiments' of Lummer & Pringsheim and Rubens & Kurlbaum had manifestly refuted the classical theory of black body radiation leading to Planck's and Einstein's superlative theories as to direct generalizations of new empirical data obtained.

However, a more punctilious analysis uncovers the following.

1. At the end of the XIX-th century, there was *no* classical theory of production and transformation of radiation in many-particle systems. For instance, the classical theory of blackbody radiation was *completely* lacking. It was advanced only *after* the solid quantum papers of Planck due to Rayleigh and Jeans's strenuous efforts and became an established one due to Hendrik Lorentz himself only in 1908. Yet Planck's pre-eminent radiation law was elaborated already in 1899, while Einstein's startling impact to the old quantum theory was made in 1905 - 1907. Incidentally, Wien's classical formula describing the dependence of density of blackbody radiation upon its frequency was merely a phenomenological one, obtained, besides, in a dubious *ad hoc* way. It had to be properly corrected by the theoretician. And Max Planck masterly applied the intricate apparatus of electrodynamics, thermodynamics, and statistical mechanics to rederive it as a sheer theoretical law and to amend Wien's half-phenomenological unrefined results.

2. On the other hand, Einstein's vehement efforts to introduce actual light quanta were not the attempts to 'save the phenomena'

and to explain away the notorious 'hard facts' that could not be explained by highly-developed grand classical theories. Although one often reads the statement that in 1905 Einstein was concerned with an explanation of the photoelectric effect, the punctilious study of the foolhardy 1905a ('light quanta') paper reveals that this was *not* the case. The measurements of the effect at that time were not sufficiently accurate to point *without any doubt* to an apparent violation of classical behavior (Ter Haar 1967). Einstein worried not so much by the evidence dealing with experiments and appealed to fluorescence, photoelectricity, and photoionization data only as to *indirect* evidence in favor of his bold light quanta hypothesis.

3. History of the old quantum theory in recent history & philosophy of science literature was ascertained to be not an unproblematic, paradigmatic case when the theory is unequivocally determined by the strong evidence. On the contrary, there was a heavy portion of theoretical indetermination there.

At first, in 1969-1984 it was convincingly exhibited (see, for instance, Boyer 1975, 1984) that most of the 'old quantum' genuine phenomena (beginning with Planck's radiation law) can be naturally fathomed in a strictly classical (Maxwellian) way in startling 'stochastic electrodynamics'. The mere presence of random classical electromagnetic zero-point radiation with a Lorentz-invariant spectrum modifies the ideas of classical electron theory to such an extent as to furnish three classical derivations of Planck's celebrated spectrum.

Secondly, Robert Hudson (1997) elucidated that, contrary to Einstein, Jeans and Poincaré reckoned (in a good company with Klein, Kuhn, and Norton) that the Rayleigh-Jeans law is *not* a law one should expect from a classical physicist, since the proper derivation of this law necessitates an odd and ad hoc equipartition hypothesis. Note that too many classical physicists (like Boltzmann) did not accept the hypothesis and respectively the classical equipartition theorem.

Hence this chapter aims to provide a rational reconstruction of the early quantum theory origin explaining the peculiarities (i)-(iii) away by exploiting the lucid theory-change model presented in Chapter I.

Viz., this chapter aims to take a decisive step towards a more complete description of the Quantum Revolution bringing some additional light on the *intertheoretical* relations between the early quantum theory and other physical theories of the day and stressing their lead in its genesis and acceptance.

In this punctilious piece of work, I have nothing to say on the relentless 'critical experiments', ultimate 'refutations', and another empiricist gobble - gooking except that these notions are irrelevant in describing an origin of the Quantum Revolution. On the other hand, the corresponding relativist notions of austere 'crisis', profound 'paradigm-shift', subtle 'gestalt switch', insurmountable 'incommensurability', etc. appear to be unnecessary too. A true origin of the second scientific revolution certainly consisted not in a clash of entrenched classical theories ('paradigms', 'research programs', etc.) with blunt 'hard facts' but in the series of stout clashes and subsequent modifications of three mellow classical theories. Experimental evidence played an auxiliary role only. Whereupon, the genuine innovations occurred because of the strenuous efforts to reconcile the three research programs of classical physics.

A few words should be uttered on the historiographical basis of the chapter. Though not completely, it is grounded on Thomas Kuhn's 1978 assiduous book and startling papers on quantum discontinuity, not to forget Rene Dugas's (1959), Martin Klein's (1962,1966), Max Jammer's (1966), Stephen Brush's (1967) and Hans Kangro's (1970) distinctive impacts. Fortunately, I am not quarreling with the facts embraced by these pre-eminent accounts. I am discontent merely with some arbitrary *interpretations* of them and especially with the skillful selection procedures picking up the 'essential' facts and disregarding the 'unimportant' ones. (For instance, I disagree with Kuhn's underestimation of Planck's impact into old quantum theory). In full accord with Imre Lakatos's (1971) salutary recipes, my rational reconstruction presents a selection of facts governed by a lucid and distinctive theory-change model lying in the heart of the matter.

2.2 Part one. An Origin of Max Planck's Assiduous Black-Body Theory.

According to Thomas S. Kuhn's superlative account (Kuhn 1978), at the end of the XIX-th century, three mature research programs were to interact (and I should add – intertwine and interpenetrate) consequentially in Max Planck's creative work: thermodynamics, electromagnetic theory, and statistical mechanics. Before 1900, he has furnished substantial contributions to all three but their weight diminished in his creativity from immaculate classical thermodynamics through Maxwell's new-fangled theory to Boltzmann's through and through controversial approach. Thermodynamics was an infallible paragon of scientific theory for him. Planck's corresponding advancements in it were well known and highly appreciated by the current scientific community before he first turned, not without substantial doubts and numerous hesitations, to queer realm of Maxwellian electrodynamics. Besides, one is not surprised by the fact that statistical mechanics entered Planck's research much later and against solid resistance. Hence, for Planck, the role of classical electrodynamics and statistical mechanics initially was purely instrumental. Maxwell's dainty equations and Boltzmann's sophisticated statistical technique provided merely conceptual tools to salve the exasperating conundrum of irreversibility first and only then the inextricable technical puzzle of blackbody radiation.

Yet it was in no means accidental that it was Max Planck who appeared at the center of the profound conflict between the basic paradigms of classical physics. Primarily, as one of the first professional theoretical physicists, Planck was particularly sensitive to the relations between different mellow physical theories. It was due to his 'sophisticated realism' philosophical standpoint concerning the structure, dynamics, and progress of physical theories. The standpoint can be properly elicited via the comparison with Mach's confounded philosophical empiricism.

According to Planck, in the very beginning of science, the division of physics was grounded on the physiological peculiarities of man and possessed, in short, sheer anthropological character. Nevertheless, while originally the basic ideas of physics were taken from the specific sense perceptions of a common man, the latter is in modern times in large measure *excluded* from physical acoustics, optics, and theory of heat. Hence,

"In short, we may say that the characteristic feature of the entire previous development of theoretical physics is a *definite elimination* from all physical ideas of the *anthropomorphic elements*" (Planck, [1910], 2015, p.4; my italics).

The peremptory result consisted in the attainment of 'unity and compactness in our system of theoretical physics'.

On the contrary, due to the positivist view, the pivotal problem of physics consists only in the reading of the sense perceptions, following experience, to fixed laws. However, retorts Planck, though self-compatible, the positivist view has never contributed to any solid advance in physics. It suffices to turn to the epoch-making writings of Copernicus, Kepler, Newton, Huygens, Faraday, et al. In a series of conspicuous lectures, eight in all, delivered with various amendments over a period ending about 1920 (and published by Dover) Planck alleged,

"Looking back over the preceding, we may say briefly that the feature of the whole development of theoretical physics, up to the present, is the *unification* of its systems which has been obtained *by* a certain elimination of the *anthropomorphous* elements, particularly the specific sense-perceptions" (Planck, [1908], 1960, p.4; my italics).

For instance, the importance of Maxwellian electrodynamics consists in the fact that it masterfully unites two branches of physics, which previously had to be treated as independent. That is why all theorems which are valid for one branch, apply to the other.

"Before the introduction of the electro-magnetic theory, physics was divided into three *separate* branches – mechanics, optics, and electrodynamics, and *the unification of these is the ultimate and greatest aim of physical research*... and thus the number of independent branches has been reduced to two – the penultimate step towards the unification of the world picture. When and how the last step will be made, the linking up of mechanics and electrodynamics, cannot be said, and though many clever physicists are at present occupied with this question, the time does not yet seem ripe for the solution (Planck [1908], 1960,p. 93; my italics;).

Consequently, Mach's celebrated epistemological theory, though extraordinary simple, clear and plain, represents, by and large, merely an understandable reaction against the proud expectations of the generations of physicists associated with special mechanical phenomena. Mach's pre-eminent principle of Economy of Thought

"*is not the nature of science* and he [Mach] evades the most convenient of all scientific research – the finding of a *fixed* world picture independent of the variation of time and people. . . .The *fixed unity* of the world picture is, however, as I have endeavored to show, the fixed goal which true science approaches through all its changes" (Planck, [1908], 1960, p.24; my italics).

That is why one should not be surprised to discover that **the red thread through all Planck's epoch-making creative work was the problem of unification** of the mature theories of the day.

"Now, how far have we advanced today toward the unification of our system of physics? The numerous independent domains of the earlier physics now appear reduced to two: mechanics and electrodynamics, or, as one may say, *the physics of material bodies* and *the physics of the ether.* The former comprehends acoustics, phenomena in material bodies, and chemical phenomena; the latter, magnetism, optics, and radiant heat. But *is this division a fundamental one?* Will it prove final? *This is a question of great consequence for the future development of physics*" (Planck, [1910], 2015, p.4; my italics).

Note that Planck himself is quite optimistic concerning the future development of physics and vehemently tries to make a substantial impact of his own.

"For myself, I believe it must be answered in the *negative,* and upon the following grounds: *mechanics and electrodynamics cannot be permanently sharply differentiated from each other.* Does the process of light emission, for example, belong to mechanics or electrodynamics?

To which domain shall be assigned the laws of motion of electrons? At first glance, one may perhaps say: to electrodynamics, since with the electrons ponderable matter does not play any role. However, let one direct his attention to the motion of free electrons in metals.

There he will find, in the study of the classical researches of H.A. Lorentz, for example, that the laws obeyed by the electrons belong to the kinetic theory of gases than to electrodynamics. In general, it appears to me that *the original differences between processes in the ether and processes in material bodies are to be considered as disappearing. Electrodynamics and mechanics are not so remarkably far apart*, as is considered to be the case by many people, who already speak of a conflict between the mechanical and electrodynamic views of the world" (Planck, [1910], 2015, p.4; my italics).

It is quite important, that, on Planck's experienced, and sophisticated view, Newtonian mechanics and Maxwellian electrodynamics are not *incommensurable*. Mechanics is grounded on the ideas of space and time, yet the same can be said about the classical electrodynamics. Hence

"In fact, there are many indications pointing toward the *ultimate amalgamation* of these two subjects, the domains of which already overlap in some measure" (Planck, [1910], 2015, p.4; my italics).

Consequently, Planck considered himself as a man who strives to provide the situation when 'the gulf between ether and matter will be once bridged' answering the question that "will characterize the whole nature of the further development of our science" (Planck, [1910], 2015, p.4).

In scrutinizing Planck's conspicuous impact, it should be punctuated from the very beginning that the so-called 'classical theory of black-body radiation' before Planck's strenuous efforts did not exist at all. And what did actually exist? – A motley hotchpotch of phenomenological and half-empirical laws either obtained at the expense of downright generalization of empirical data or due to physically unfathomable and ad hoc premises doggedly inserted into the body of knowledge. For instance, in 1879 Josef Stefan, as an extrapolation from preliminary experiments, demonstrated that the dependence of the blackbody radiation density u upon temperature T is displayed by $u = \sigma T^4$.

Hereafter Langley, W. Michelson, H. Weber, F. Paschen, et al. had enunciated a bunch of manifold expressions for blackbody distribution function u (ν, T), doggedly derived from the experiments, until

Wilhelm Wien came to grips with the problem. He had pioneered in furnishing the expression for u (ν, T) from purely *theoretical* considerations yet dubiously and speculatively. For Wien (1896), a heated gas constituted the source of blackbody radiation. Following the perspicacious line of thought of a Russian physicist Vladimir Michelson, Wilhelm Wien elicited that, in the gas, number of molecules with velocities in the range between v and v + dv is, by Maxwell's distribution law, proportional to ν^2 exp (-ν^2 /α^2) with α proportional to the gas temperature T. If one makes the preternatural assumption that both the frequency and the intensity of the radiation from a certain molecule are functions *only* of the molecule's velocity, then the distribution of radiation obeys the foolproof equation

U (ν, T) = F (ν) exp (- f(ν/T).

On determining F (ν) and f (ν), Wien arrived at the following odd expression (in modern notation)

u (ν, T) = 8πh / exp (hν/kT).

Clearly, as a purely theoretical product, the Wien half-phenomenological distribution law had little authority. However, the situation had changed gradually, when Max Planck rederived it by a remarkably diverse route in 1899. His strenuous efforts transformed a shabby empirical law obtained in an ad hoc way into a sheer theoretical one, with the domain of validity established: hν/ kT $\gg$ 1. Hence, Planck's assiduous papers turned out to be the first actual and successful effort to contrive a trailblazing and solid theory of blackbody radiation, i.e. to rederive the radiation density u (ν, T) from the 'first principles' of classical mechanics, thermodynamics and electrodynamics as opposed to derisive phenomenological efforts to guess the right expression through multifarious and controversial experimental results.

Surely, Planck met many exasperating rubs on the way, punctiliously deriving complicated intermediate theoretical laws, comparing them with intricate experimental results, sedulously correcting the laws, and so on. He famously was an immaculate theoretician par excellence with thermodynamics taken as a paragon of scientific rationality. It was in his times when thermodynamics became established, carefully thought and composed, highly respectable physical theory with a throng of practical applications. When he came to a frustrating conclusion that thermodynamics is insufficient, he turned to controversial Maxwellian electrodynamics – young, unfinished, and dubious at those times, whose proper empirical corroboration had just commenced primarily due to Hertz's strenuous efforts. Nevertheless, Planck, in his unending quest for irreversibility, had to insert imaginary vibrating resonators into his primary theoretical scheme; they were represented by tiny oscillating currents governed by Maxwell's equations. Planck's dainty oscillators had nothing to do with real experimental devices. They had nothing to do with actual molecules, atoms, or even electrons representing merely whimsical theoretical devices for establishing thermodynamic equilibrium between matter and radiation.

In 1897, Planck published the first paper of a five-part series humbly entitled '*On Irreversible Radiation Processes*' (Planck, 1897a). In all the five papers (Planck, 1897, a,b,c,1898,1899), as well as in the think piece, which recapitulated their results for the '*Annalen der Physik*' (Planck, 1900b), he thoroughly inquired into the properties of the ideal model embracing a system of resonators interacting with an electromagnetic field. However, after he had read his first paper to the Academy, a sharp and probative critique by Boltzmann unexpectedly was laid out. Boltzmann convincingly disputed that, though Planck's fine expressions for resonator absorption and emission were correct, the 'irreversibility' research program for which they were punctiliously designed should ineradicably fail. Both Maxwell's dainty equations and the trenchant boundary conditions on their solution are ineradicably invariant under time reversal.

When the indispensable application of Maxwell's equations transpired to be insufficient, Planck had to apply statistics – quite unwillingly, step by step, though under the incessant pressure of experimental results. Yet the following startling story is of particular importance here (Dugas 1959; De Broglie 1962, p.139). Planck, who vehemently eagered to broaden the realm of the validity of statistical thermodynamics, advanced (with a help of classical continuous notions) sophisticated thermodynamics of electromagnetic radiation and strived to intrude the entropy of radiation by analogy with its energy. Being an admirer of pre-eminent Ludwig Boltzmann, Max Planck conveyed to the 'father of statistical mechanics' the results of his multifarious investigations, and presented one of his deft papers to Boltzmann's strict judgment. However, Boltzmann bluntly retorted that Planck would never be able to create a solid theory of statistical thermodynamics of radiation without an offhand inserting into the tissue of the theory of formidable and previously unknown *element of discontinuity* into processes of radiation.

Even in 1872, 28 years before Planck's 1900b masterpiece, Boltzmann in his startling paper *'Further investigations of thermal equilibrium between gas molecules'* tentatively applied the bizarre notions of discrete energy to the exchange processes. Deriving the second law of thermodynamics from a statistical point of view, Boltzmann gradually arrived at the daring notion of *'energy atoms'* in the complicated processes of molecular interactions. The finite energy amounts that can be freely exchanged for colliding molecules lead Boltzmann to important calculations of the number of collisions with the help of specific 'combinatorials'. However, due to the current dominant Weltanschauung (worldview), Boltzmann took the notions of energy quanta with discretion just as an auxiliary mathematical subterfuge. Incidentally, 'Boltzmann's heritage' was rather controversial and jolly debatable. At first, Planck strived to get through simply by conjuring up an electromagnetic analog of Boltzmann's H-theorem. When it turned out to be insufficient, he had to utilize the exasperating probability calculus and the notorious combinatorial definition of entropy, which he whole-heartedly disliked. Thus,

"By mid-winter 1897-98, at the latest, Planck was studying Boltzmann's version of the second law with care, was exploiting suggestions he found there, and had abandoned or *all but abandoned his resistance to Boltzmann's approach.* Unfortunately for historians, he did not explicitly acknowledge his change of mind for almost two years, a delay that has reinforced the almost universal impression that *his conversion to a statistical viewpoint was intimately associated with his introduction of a quantum hypothesis* at the end of 1900" (Kuhn 1978, p.78; my italics).

However, quite the opposite is true: *Planck's thrilling introduction of the bizarre quantum hypothesis in 1900 constitutes a direct consequence of his indispensable yet reluctant conversion to a statistical viewpoint!* Quite fortunately, one has *direct* evidence of Planck after the events. In a blunt letter to Robert Wood (Planck 1931), he admitted that

"Boltzmann explained the existence of thermodynamic equilibrium through statistical equilibrium; if his considerations are applied to equilibrium between matter and radiation, one concludes that the transformation of all the energy into radiation [demanded by classical physics] can be avoided by the supposition that energy should exist from the very beginning in some *discrete portions*" (my italics).

Planck many a time and oft recalled that all his hard life, in publications or private correspondence, Boltzmann was stressed cold with him. Only in his last years, when Planck informed him about the stout atomistic justification of his radiation law, Boltzmann changed his attitude radically and became jolly friendly to Planck.

By the beginning of 1900, only one facet of Boltzmann's masterful treatment of irreversibility was still absent from Planck's far-going scientific program, viz. the use of combinatorials, and by the end of the year, Planck took that step, too. What impelled him to do so was no longer the involved problem of irreversibility. It was rather the incessant search for a radiation law that could ultimately pass the test of new, more refined experiments.

The apparent imperfection of the derivation of the Wien distribution law, that Planck submitted to the '*Annalen* ' in November 1899, consisted in the lack of a uniqueness proof for the function he had shrewdly defined as oscillator entropy. Pointing on their measurements of the frequency distribution of radiation from a novel piece of fine experimental apparatus, laboratory black cavity, O. Lummer, and E. Pringsheim rendered a new formula for the blackbody radiation that deviated from the abovementioned Wien's law. Planck closed out the obstacle in a paper submitted for publication already in February 1900. In that paper, he alleged to have derived, rather than defined, oscillator entropy for the first time, supporting Wien's law once more.

Yet the experiment proved stubborn. In a paper, reported to the Physics Section of the Naturforscherversammlung on 18 September

1900, Lummer and Pringsheim retorted that the Wien-Planck stale distribution law did not assent with their refined measurements on black radiation in the low-frequency domain. In this range, achieved only with the aid of recently advanced techniques, the discrepancies between experiment and theory were near 50% and evidently could not be due to common experimental errors.

The evidence was ineradicably convincing, but Planck was ready to meet it now. In a deft paper presented to the Physical Society on 19 October, he referred to the recent proof of the Wien law he had submitted to the '*Annalen der Physik*' in March, and at once pointed its drawback. The entropy of n oscillators must, he enunciated, depend not simply on their total energy, as was surmised, but on the energy U of a single oscillator.

The expression $\delta^2 S/\delta^2 U = -\alpha/U$ earlier found is too blunt and should be exchanged for a more intricate term. Planck had elaborated an equation that 'is the simplest by far of all the expressions which yield S as a logarithmic function of U (a condition which probability theory suggests) and which besides coincides with the Wien law for small values of U'. If the equation for S is taken as the first term $(-U/\alpha)$ in a power series expansion of $(\delta^2 S/\delta^2 U)^{-1}$, his new form follows immediately by the addition of a term proportional to U^2. With $U = -\alpha/U (U+\beta)$, two integrations, the standard condition $\delta S/\delta U = 1/T$ and an application of the displacement law yield a novel thrilling distribution law

$$U = b\nu/exp(\alpha\nu/T) - 1.$$

This queer radiation formula, Planck maintained,

"so far as I can see by quick inspection, represents the hitherto published observational data just as satisfactorily as the best previously proposed distribution function...I, therefore, feel justified in directing attention to this new formula, which, from the standpoint of electromagnetic radiation theory, I take to be the simplest excepting Wien's".

New refined measurements quickly demonstrated the last equation to be undoubtedly superior to all the other distribution laws. Nevertheless, Planck had to discover the route to the formula that was less ad hoc.

"On the exact day, Planck recalls, when I first formulated this law, I began to devote myself to the task of investing it with *real physical meaning*, and that issue led me of itself to the consideration of the relationship between entropy and probability, and thus to *Boltzmann's [distinctive] line of thought*" (Planck 1909; my italics).

As Thomas Kuhn perspicaciously punctuated, those brief remarks have regularly been awry fathomed as reflecting Planck's initial conversion from phenomenological thermodynamics to statistical one. However, that striking turn-about took its place at least a year, and more probably, three years *before*. When Planck referred to the renowned 'relation between entropy and probability', he did not have in mind the grand statistical approach in general but only, as his words humbly suggested, Boltzmann's intricate combinatorial definition of entropy. Planck, who must have discovered this definition in Boltzmann's immaculate "*Gas Theory*", turns out to have been the first scholar to acknowledge even its existence.

Alas, Planck's initial derivation of the Wien law contained an annoying internal contradiction. The n resonators he comprehended were required to be *independent*, but his subtle argument depended on surmising that their total energy Un was distributed *equally* among them. A sophisticated argument would reveal the various ways in which that energy might be divided between resonators just as Boltzmann himself, in the corresponding combinatorial arguments, had resolutely divided the total energy of gas among its numerous molecules. The expression for Planck's distribution law $U = b\nu / \exp(\alpha\nu/T) - 1$ can be manipulated to yield $1/T$ as a function of U and ν, and $1/T$ is just $\delta S/\delta U$. After integration one decidedly arrives at

$$S = (b/a) \log \{(1 + U/b\nu)^{1+U/b\nu} / U/b\nu^{U/b\nu}\} + \text{const}$$

Planck seemingly should have been strongly encouraged by its evident resemblance to Boltzmann's pre-eminent expression for the logarithmic relation between entropy S and probability W. Yet the thrilling equation seems to apply only to a single resonator with energy U in equilibrium with a radiation field; it is not suitable for proper interpretation in probabilistic terms.

So let one just imagine N independent resonators of frequency ν in thermal equilibrium with their radiation field. Their total entropy is necessarily equal to NS, while their total energy is equal to NU. If the abovementioned exasperating combinatorials are to be implicated, the total energy must be subdivided into P elements of equal size ϵ, so that $P\epsilon=NU$. Multiplying Planck's equation for S by N *and substituting* $P\epsilon/N$ for U definitely yields

$$SN = (b/a) \log \{(N + P\epsilon/b\nu)^{(N+P\epsilon/b\nu)} /N^N (P\epsilon/b\nu)^{(P\epsilon/b\nu)} \}+ const$$

To arrive at the expression involving only integers, the size of the energy element ϵ must be set equal to $b\nu$. The quantity in {}then reduces, for large N and P, to (N+P-1)!/ (N-1)!P! However, that queer expression is the standard one for the well-known number of ways in which P indistinguishable elements can be distributed over N distinguishable boxes. The first stages in Planck's utilization of Boltzmann's renowned relation between entropy and probability were successfully terminated.

However, the intricate problem remained. The queer combinatorial expression elicited by working backward from Planck's distribution law was decidedly *different* from the one Boltzmann had unfolded in deriving the equilibrium distribution of gas molecules. Hence, Planck had to exhibit that it is proportional to the probability appropriate to equilibrium radiation. Planck's ultimate problem was to compute the entropy S of a particular distribution of the total energy E over N resonators. Then he had to find its maximum concerning the variation of the distribution of the total energy E over frequency ν. Yet to compute the entropy of an arbitrary distribution, Planck had to implicate combinatorials and hence he punctiliously followed Boltzmann in subdividing the energy continuum into tiny elements of finite size.

"We must now give the distribution of energy over the separate resonators of each group, first of all, the distribution of the energy E over the N resonators of frequency ν. If E is considered a continuously divisible quantity, this distribution is possible in infinite ways. We consider, however, – this is *the most essential point* in the whole calculation– E to be composed of a very definite number of equal parts and use there the constant of nature h = 6,55 x 10-27 (erg x sec). This constant multiplied by the frequency, ν, of the resonator yields the energy element ϵ in ergs, and, dividing E by ϵ, one obtains the number,

P, *of energy elements to be distributed over the N resonators*" (Planck 1900b; translated by Ter Haar 1967, p.83; my italics).

Then Planck defines a '*complexion*' (an expression, he punctuates, employed by Boltzmann himself for a similar concept) as a particular specification of the set of numbers, which fixes the number of elements ϵ attributed to the various resonators in the set of N. The total number of possible complexions is R. To get the equilibrium distribution one has to maximize R or $logR$ by varying the energies at the various frequencies. Straightforward calculations demonstrate that the entropy (log R) will have a maximum if

$U\nu \quad = h\nu/ \exp (h\nu/kT) - 1.$

The corresponding distribution for the field is

$u_\nu = (8\pi^2\nu^2) U_\nu.$

Incidentally, both in his original derivation papers and far more apparent, in his fine 'Lectures' (1906) as well, Planck's deft radiation theory is incompatible with the quantization of the resonator energy. The theory merely requires fixing the size of the small intervals into which the energy continuum is subdivided for purposes of subtle combinatorial computation. In Planck's prudent theory, resonator emission and absorption are governed in full by Maxwell's equations. Indeed, Planck did commonly write expressions like UN = Phν. Yet the UN is the *total* energy of N resonators. Restricting it to integral multiples of hν does not impose any similar restriction on the energy of an *individual* resonator, which may still vary continuously. Just as Planck later acknowledged in his immaculate "*Eight Lectures on Theoretical Physics*" first delivered at Columbia University in 1909 and published in 1915 by Columbia University Press in New York,

"J.J. Thompson inclines to the most radical view, as do J. Larmor, A. Einstein, and with him I. Stark, who even believe that the propagation of electromagnetic waves in a pure vacuum does not occur precisely in accordance with the Maxwellian field equations, but in definite energy quanta hν. I am of the opinion, on the other hand, that at present it is *not necessary to proceed in so revolutionary a manner*" (Planck [1910], 2015, p.95; my italics).

Thus, though Planck constantly appealed to the experimental results, the role of the experiment should not be exaggerated. Blackbody experiments played the role of peremptory factor, which impelled Planck to apply statistics in the growing rates. In the lack of experimental data, Planck would not use Boltzmann's notorious combinatorials at the full rate since he disliked them whole-heartedly. The latter was grounded on introducing distinctive and dubious hypotheses that Planck, an admirer of classical thermodynamics, incessantly tried to avoid.

2.3 Part two. A Brief Discussion: Max Planck vs James Jeans.

The role of experiments in constructing the quantum theory was trenchantly outlined by Planck himself:

"Even today, physics can be divided into three *essentially different* groups of theories: *mechanics*, including elasticity, hydro-dynamics, and acoustics, then *electro-dynamics*, with magnetism and optics, and *thermo-dynamics*. Each of these three groups of theories has retained a certain degree of independence, though there are to-day a large number of points of contact between them, either supplementing or clashing with each other. *Thanks to the rapid advances of experimental science,* the number of these points of contact is continually increasing" (Planck, 1960, p. 83; my italics).

Yet one cannot maintain that Planck's distribution law was a direct generalization of experimental results. On the contrary, it was all upside down. Planck's fatiguing route to it was from top to bottom. He had to take the experimental evidence into account, yet it impelled him not to assiduously 'deduce from phenomena' but to apply more theory instead. As one of the first professional theoreticians in physics, an eminent ' der Patriarch' of German theoretical physics, Planck can in no ways be described as a snotty boy lucky to find a law of Nature the value of which he could not comprehend. As a professional theoretician, Planck was extremely sensitive to the importance of the problem he strived to solve and to the emergence of scrutinizing it in the *intertheoretical context*. He conceived that the origin of the problem lies in the profound cross-contradictions between classical mechanics, statistics, electrodynamics, and thermodynamics (Planck 1910, p.616).

Besides, more apparently in startling 1906 '*Lectures on Heat Radiation*', Planck applies essentially the same arguments as Einstein in his 1905a daring paper:

"Introduction of probabilistic notions into the electromagnetic theory of heat radiation means the intrusion of an entirely new element that is substantially *aligne*d to the fundamentals of electrodynamics. Hence, from the very commencement, the principal question occurs on the justification and real necessity of such notions. At first sight, one can easily conclude that theory of probability has no place in the theory of pure electrodynamics. Indeed, electromagnetic field equations together with initial and boundary conditions are known to determine the duration of an electromagnetic process in time. Hence, all the considerations disconnected with field equations seem to be substantially unjustified and at least not necessary. Either they lead to the same results given by the main equations of electrodynamics, – and hence they are needless, - or they lead to other results and hence they are false.

Despite the unsolvable *dilemma,* there is a cleavage in the above considerations. Indeed, digging still deeper into the heart of the matter, one can ascertain that both 'initial and boundary conditions' and 'time duration' of a process, as we comprehend them in electrodynamics, are not the same in thermodynamics. To conceive the problem situation better one has to turn to a concrete example considered in the last chapter – to hohlraum radiation that is homogeneous in all directions. From a thermodynamic standpoint, the state of radiation is entirely determined, if the intensity J_ν of the monochromatic radiation for all the frequencies ν is given. For electrodynamics, all these are insufficient; from this standpoint to determine the state one has to determine each of six magnetic and electric components of the field in all the space points..." (Planck 1906, p.105; my italics).

It was the fact of startling origin of 'the old quantum theory' from the clash between classical electrodynamics and statistical mechanics that was indicated by one of the leading Russian theorists of the XX-th century beginning, a Professor Emeritus of Kazan University Dmitry

Goldhammer (Goldhammer 1911; see the epigraph).

The seminal contributions of Michelson, Wien, Boltzmann, Planck, and Einstein constituted the milestones of deep penetration of statistical methods and concepts into radiation theory, the milestones of regarding the electromagnetic field from a statistical point of view (Bucklaev 1957). According to Planck's own words,

"However, it appears that this [i.e. the modification in mechanics due to the principle of relativity consisting in the introduction of a new universal constant c entirely foreign to classical mechanics] is not the only, nor the most difficult sacrifice which dynamics must offer if it wishes completely to embrace the theory of heat. It has probably not finished with the question of the discontinuity of matter. The laws of heat radiation, specific heat, electron emission, of radio-activity, and yet many other branches are in agreement that not only matter itself but also the effects radiated from matter ... possess discontinuous properties, which again can be characterized by a new natural constant: the elementary quantum of action" (Planck [1908],1960, p.87).

In a sense, one can profess the existence of a mellow Boltzmann-Wien-Planck synthetic research program of reconciling Maxwell's electrodynamics and Boltzmann's statistical mechanics. (See Chapter I on synthetic and reductionist research programs).Subsequent versions of the wayward program can be characterized by ad hoc and non ad hoc reactions to the experimental results. Note that the program provided a constant empirically progressive problem shift in comparison to its rival – the imperceptible reductionist program of James Jeans.

Incidentally, the primary ideal model of the abovementioned synthetic program was pioneered in the second half of the XIX-th century by Russian physicist Vladimir Michelson. He sought to explain a distinctive analogy between the form of the experimental curve describing the frequency dependence of blackbody radiation energy and Maxwell's eminent distribution curve in the common theory of gases. Though Wien advanced the Michelson's ideal model, the hypotheses that provided the turnabout from the initial version of the program to the next one, transpired so ad hoc, that the first sheer theoretical derivation of the blackbody radiation spectrum in the high-frequency limit was provided only by Max Planck.

However, the accommodation of these theories on the crossbred level impelled the corresponding changes in the foundations of the theories in the future. In particular, the abovementioned inconsistency in the energy distribution between oscillators further brought to the quantization of individual oscillator energy.

The irrefutable 'hard core' of the synthetic program was outlined in the long quotation given (Planck 1906) and consisted of three constructively independent mellow theories: Maxwellian electrodynamics, classical thermodynamics, and Boltzmann's statistical mechanics. The 'positive heuristic', as opposed to the Jeans' reductionist program, consisted in the assumption that the electromagnetic radiation in some aspects shows corpuscular properties up and hence can be described by the laws, concepts, principles, and methods of statistical mechanics. The 'protecting belt of auxiliary hypotheses' was contrived by a system of models M_1, M_2, ..., M_k that rendered more and more exact 'scientific pictures of reality'. Some of the resulting theories were regrettably ad hoc but were later exchanged for the better ones that corrected their offhand faults. For instance, some of Planck's deft theories were apparently ad hoc, and his distribution law was yielded in an ad hoc manner first but was properly corrected soon.

The hard core of the Jeans blunt reductionist program was made up of Maxwellian electrodynamics. The positive heuristic of the program consisted in the assumption that the physical situations ascertained in the blackbody punctilious experiments are not the cases of equilibrium at all. Hence, the negative heuristic manifests that thermodynamics and statistical mechanics are inapplicable in these cases. The protecting belt of the program represented a sheer subterfuge capable to wriggle from the 'ultraviolet catastrophe': millions of years, eternity might be required to transmit energy from the lower modes of vibration to higher ones. A true equilibrium might never be achieved. Hence, the notorious Maxwell-Boltzmann equipartition theorem is inapplicable.

The history of subsequent competition of the synthetic program with its artful rivals, especially with the reductionist program of Jeans, is fascinating in itself. We will not pay much attention to it here. The main trouble with Jeans' sordid program consisted in that not only the fine experiments on the distribution of radiant energy became relentlessly inexplicable. By denying that these experiments dealt with equilibrium situations, Jeans simultaneously denied the relevance of immaculate thermodynamic arguments to them. Moreover, novel derivations would have to be found for Kirchhoff's law, the Stefan-Boltzmann law, the Wien displacement law, and so on. Though Jeans furnished a stimulating and skillful explanation of the Stefan-Boltzmann law within his willful program, it was quickly supplanted by the quantum program with its pre-eminent successes in the purview of specific heats.

Contrary to Jeans' sordid venture, the prosperous quantum synthetic program willingly procured the implantation of the crossbreed objects constructed from the basic objects of Maxwellian electrodynamics, classical thermodynamics, and statistical mechanics. All the brave new models facilely constructed within the synthetic program were crossbred ones but in different extents. Incidentally, they became crossbred more and more under the influence of the abovementioned experiments.

Planck many a time and oft sagaciously noticed that the intrusion of $E = h\nu$ was caused by a profound fissure between statistical mechanics and electrodynamics. Though the ultimate elimination of the cross-contradiction consisted in the creation of quantum electrodynamics, i.e. in the construction of the stupendous quantum theory of radiation that took electromagnetic particles as Boltzmann's molecules that can gain energy under collisions with usual molecules and resonators (Landau 1958). The second man who became aware of the spellbinding chasm and tried to take the next steps in salving the profound contradiction was Albert Einstein.

3 AN ACME OF THE SECOND SCIENTIFIC REVOLUTION: HOW EINSTEIN'S 1905 FOOLHARDY LIGHT QUANTA GOADED DISCREET SPECIAL RELATIVITY

Fundamental crisis set in, the seriousness of which was suddenly recognized due to Max Planck's investigations into heat radiation (1900). The history of this event is all the more remarkable because, at least in its first phase, it was not in any way influenced by any surprising discoveries of an experimental nature.

Albert Einstein

In the relativity theory, it is no question of a revolutionary act but of natural development of the lines, which have been followed for centuries.

Albert Einstein

Abstract. An acme of the second scientific revolution was made up by Einstein's 'annus mirabilis'. To fathom in what felicitous ways Einstein's multifarious 1905 writings hang together one is bound to pay special tribute to his longed, renowned strive for unity evinced in incessant attempts to coordinate the ingrained classical research traditions.

Einstein's vehement efforts sprung out of Max Planck's pioneering attempts to comprehend electromagnetic phenomena through the lenses of conceptual structures of thermodynamics and statistical mechanics. It was Planck, who realized the cross-contradiction between 'the physics of material bodies' and 'the physics of the ether' and outlined the first sketch of its withdrawal: *the paradigms 'must be modified to remain compatible'*. And it was Planck who took the first step in modifying the physics of the ether and contending that 'not only matter itself but also the effects radiated from matter possess discontinuous properties, which again can be characterized by a new natural constant: the elementary quantum of action'. Einstein's part consisted in that he took the next step in modifying the second component – the physics of material bodies. 'The modification in mechanics due to the principle of relativity is the introduction of a new universal constant, entirely foreign to classical mechanics, namely the velocity of light in an absolute vacuum' (Planck). Einstein's foolhardy light quanta hypothesis and distinctive special theory of relativity turn out to be mere milestones of the unwinding of Maxwellian electrodynamics and statistical thermodynamics reconcilement research program. The notorious conception of luminiferous ether transpired to be a substantial impediment for Einstein's wayward statistical thermodynamics in which flagrant light quanta paper played the pivotal lead. Herewith Einstein was aware that his enticing light quanta hypothesis was too audacious to be taken literally. Hence he laid out the indispensable 'electrodynamics of moving bodies' in a markedly Duhemian phenomenological way and the basic relativity postulate was disposed as originating from Machian unfathomable 'instinctive knowledge 'doctrine. In his relentless criticism of the moldy research traditions of classical physics, Einstein was evidently influenced by David Hume and Ernst Mach. Nevertheless, when related to *creative* momenta, his 1905 reconcilement *modus operandi* was initiated by Mach's pre-eminent principle of economy of thought but taken in the subtle context of Stevin-Mach wayward doctrine and with promising inspirations of the regulative spirit of Kantian and Duhemian epistemologies. It is alleged that for Einstein Duhem's confounded conventionalism was an indispensable bridge connecting and accommodating two quite disparate subjects of Machian crude empiricism and Kantian academic apriorism and transcendentalism and constituting an indispensable mean of saving special relativity whimsical postulates from scarce criticism for their inseverable connection with the foolhardy ideas of light quanta.

Keywords: scientific revolution, Planck, Einstein, statistical thermodynamics, light quanta, special relativity, Stevinus, instinctive knowledge, Mach, economy of thought, Duhem, Kant, regulative principles.

3.1 A Brief Introduction: Einstein's Unification Attempts Reconsidered.

It is as clear as a pikestaff that Albert Einstein's multifarious writings were incessantly motivated by lofty Ideal of Unity (see, for instance, van Dongen's thorough 2010 book and numerous references cited therein). It is no wonder that in ample Schilpp volume Einstein, recollecting his renowned achievements, plainly professed that

"The special aim which I have *constantly* kept before me is *logical unification* in the field of physics" (Einstein 1949a, p. 400; my italics).

Respectively, the dazzling scientific career of Einstein even after 1915, i.e. after the general theory of relativity (GTR) had been successfully contrived, presented a downright, longing and incessant search for unitary theories: immaculate Cartan-Weyl geometrical 'Theory of Everything', Kaluza inspired enticing five-dimensional amalgamation of gravitation and electrodynamics, and so on. It goes without saying that the intrepid Quest for Unity of Nature is best laid out definitely by *these* celebrated attempts of Einstein towards unitary theories than by his early distinctive writings.

However, in my judgment, Einstein's mellow unification attempts and especially his dainty GTR grew out of his early writings and first and foremost out of his 1905 dizzy attempts to create special theory of relativity (STR), as well as out of his daring 1905 light quanta divination. For example, as Einstein repeatedly recalled later (see, for instance, Pais 1982), his perseverant efforts to set up the basic GTR tenet – the renowned principle of equivalence – were necessarily drawn upon his quite recent experience of putting up the STR.

Likewise, his foolhardy 1905 light quanta paper commences with displaying

"a *profound formal difference* between the theoretical conceptions physicists have formed about gases and other ponderable bodies and Maxwell's theory of electromagnetic processes in so-called empty space" (Einstein 1905a, p.86, my italics).

The 1905a paper ultimately aims at reconcilement of the profound research traditions of classical physics such as statistical mechanics and Maxwellian electrodynamics. Similarly, Einstein's STR paper also famously starts with eliciting a "*deep asymmetry*" (Einstein 1905d, p.140) in the common description of electromagnetic induction.Withal, Einstein certainly was not the first in underscoring the cross-contradiction between the primary paradigms of classical physics. At the end of the XIX-th century, three stout classical research traditions were to intertwine relentlessly in Max Planck's inextricable theory of quanta: thermodynamics, electromagnetic theory, and statistical mechanics (Kuhn 1978). Ere 1900, Planck has provided conspicuous contributions to all three. One of the first sheer theoreticians in physics, a recognized leader of a world-famous German school, he was particularly sensitive to the subtle intertheoretical *context* of the foundational problems of *fin de ciecle* science (see his startling lectures reprinted in Planck 1960 for convincing details). He trenchantly comprehended an origin of the host of foundational problems lying in the profound lingering cleavage between mechanics, statistics, electrodynamics, and thermodynamics. In the first one of the eight epoch-making lectures on theoretical physics delivered at Columbia University in 1909 (translated by A.P. Willis and first published in 1915 by Columbia University Press) Max Planck shrewdly alleged:

"The numerous, independent domains of the earlier physics now appear reduced to two: mechanics and electrodynamics, or, as one may say: the physics of material bodies and the physics of the ether. The former comprehends acoustics, phenomena in material bodies, and chemical phenomena; the latter magnetism, optics, and radiant heat. But *is this division a fundamental one? Will it prove final? This is a question of great consequence for the future development of physics.* For myself, I believe it must be answered in the negative, and upon the following grounds: *mechanics and electrodynamics cannot be permanently sharply differentiated from each other.* Does the process of light emission, for example, belong to mechanics or electrodynamics? "(Planck, 1910, pp.4-5, my italics; see also Planck's 1906 immaculate lectures on the theory of heat).

Hence, *the overall aim of the present chapter is* to take the next step and to reveal the abiding influence of the abovementioned reconcilement on Einstein's 1902-1905 startling writings on quantum theory & STR advancement. Respectively, the second part of this chapter deals with the realm of reconciliation problems that irrepressibly brought Einstein to the peculiar domain of 'electrodynamics of moving bodies'. Correspondingly, the third part aims at answering the alluring question: what was a prodigious train of thought that engendered Einstein to contrive light quanta and STR on the solid basis of Planck's promising investigations. It is contended that the former and the latter turn out to be only milestones of implementation of Maxwellian electrodynamics; statistical mechanics and thermodynamics productive reconciliation program that originated due to Planck's pioneering efforts. The pivotal lead in the research program was played by Einstein's 1905a light quanta surmise, since it was definitely the notorious ether conception that substantially *hindered* the realization of Einstein's daring statistical-thermodynamics design. In my view, Einstein was aware that his inscrutable light quanta hypothesis was obviously audacious and that he needed to be especially discreet; hence, he cautiously showed off his distinctive 'electrodynamics of moving bodies' in a markedly Machian / Duhemian phenomenological way. In particular, Einstein laid out his basic relativity postulate as originating from Machian 'instinctive knowledge'. Hence, finally, I will aim to emphasize once again that the pivotal concept necessary to fathom Einstein's 1905 diverse writings *as a whole*, as well as the subtle *order* of their arrangement certainly is pre-eminent Mach's principle of Economy of Thought. Yet I insist that it should be indispensably taken in the context of the Stevinus-Mach 'instinctive knowledge' doctrine and with some budding inclinations of Kantian epistemology presuming the coincidence of both constructing theory and integrating intuition of Principle, as well as Duhem's daring conventionalism.

3.2   Part one.   Young Albert and the Boffins: Peremptory Influence of Webers, Helmholtz, Hertz, Poincaré, Hume and Mach.

In good old Germany of the second half of the XIX-th century, Maxwell's ardent efforts to concord the research programs of Young-Fresnél, Faraday, and Ampére-Weber (Nugayev 2015) were stiffly reinforced by Hermann Helmholtz. In Helmholtz's judicious paradigm, charges and currents were taken as the ultimate sources of electrical and magnetic fields. Eventually, it led to Hendrik Lorentz's dainty dualistic worldview of the field equations and the equations of motion exhibited in his 1892-1900 eminent writings. Not by chance, Lorentz's theory was an artful fusion of Maxwell's decuman field theory and Wilhelm Weber's deft particle theory of pre-Maxwellian electrodynamics.

Besides, it was young Albert Einstein who dared to come to grips with the problem after such luminaries as Maxwell, Helmholtz, and Lorentz. On 10 August 1899 'Paradies' hotel letter an ambitious ETH (Eidgenossiche Technische Hochschule) student confessed to his fiancée and associate that

"I am more and more convinced that the electrodynamics of moving bodies, as presented today, is not correct, and that it should be possible to present it in a *simpler* way. The introduction of the term 'ether' into the theories of electricity led to the notion of a medium of whose motion one can speak without being able, I believe, to associate a physical meaning with this statement. I think that the electric forces can be directly defined only for empty space [...] Electrodynamics would then be the theory of the motion of moving electricities and magnetisms in free space: which of the two conceptions must be chosen will have to be revealed by radiation experiments" (Doc. №51 of Einstein 1987, p.131; my italics).

From the very start of his distinctive scientific career Einstein, seemingly goaded by Mach, had many a time and oft expressed downright doubts on the role of 'des Namens Aether' in classical electrodynamics. Yet his superficial skepticism was selectively directed at Hertz's peculiar concept of the ether as *a medium with a certain state of motion*, but not at the ether concept *itself*. It was because Einstein attributed major significance to the solid concept of 'Elektrische Massen' and regarded electric currents as *actual* motions of such charges in empty space, and not as tangled 'Verschwinden Elektrische Polarisation in der Zeit'. At the start of Einstein's dizzying scientific career his distinctive *Weltanschauung* was grounded for obvious reasons on the trenchant lectures on electricity of his ETH physics professor H.F. Weber, as it is unfolded by Einstein's lecture notes (see Doc. 37 and the related comments in Einstein 1987, pp. 223-225).

The 'substantive' concept of electricity was diligently contrived by Wilhelm Weber and was widely endorsed by an influential community of German-speaking physicists, including H.F. Weber. Therein, at least initially, Einstein's common views on electrical masses moving in the immobile ether were *quite similar* to the well-known dualistic paradigm of H.A. Lorentz. Respectively, young Einstein accomplished the abovementioned letter recapitulating that 'Strahlungversuche' was indispensable for choosing between the two standpoints he outlined. And his subsequent, 10 September 1899 'Paradise', letter to Mileva considered a sound idea for experimentally ascertaining the influence of motion relative to the ether on the propagation of light in transparent bodies.

Nevertheless, Einstein's venerable professor of physics expressed no explicit enthusiasm for his work, and Albert made no further allusions in his papers and correspondence of his setting force the electrodynamics of moving bodies *for almost two years*. Nevertheless, 'die prinzipielle Trennung von Lichtaether und Materie', 'Definition absoluter Ruhe', and so on surely were among the multifarious topics he vividly conferred with his close friend Michele Besso (see especially Einstein's 4 April 1901 letter to Marić). In March 1901, Einstein confidentially conveyed to Mileva that he looked forward to the perfection of 'unsere *Arbeit uber die Relativbewegung*'. In September 1901, he proudly informed his close friend and future collaborator Marcel Grossman on working up a finer method for the examination of the motion of matter relative to the ether, grounded on 'gewonlichen Interferenzversuchen'. Yet by December 1901, he was 'arbeite eifrigst' on exasperating '*die Elektrodynamik bewegter Körper*', that promised to become "eine kapitale Abhandlung" (Einstein's 17 December 1901 letter to Marić). A sad calculation setback had earlier impelled him to renounce his 'Ideen über die Relativbewegung', but now he believed in these farsighted ideas even more. He confided the motley stuff to prof. Kleiner and the latter even

"thought that the experimental method proposed by me is the simplest and most appropriate and conceivable. I was very pleased with the success. I shall certainly write the paper in the coming weeks"

(Einstein's letter to Marić, 19 December 1901, p. 189).

Despite prof. Kleiner's magnanimous encouragement and Einstein's young enthusiasm, *no publication on this subject* ensued for over three years – until 21 June 1905. - Why? What was behind all these? - Einstein was actually delved into plying on a 'capital treatise' on the electrodynamics of moving bodies at the end of 1901. Then he had ceased and retraced to impudent manuscript only in 1905. What happened at that time, and why had young Albert, being initially a *moderate supporter of* the ether, become its *committed foe?*

To elicit a sober reason one has first to recall mature Einstein's derogative evaluation of his youth writings – 'my worthless beginner papers' (Einstein / Marić 1992). The evidence at hand uncovers that the planned solid *'kapitale Abhandlung'* was a 'far cry' from the dainty STR paper. On the contrary, now one knows without a doubt (allegedly due to Rynasiewicz 2000) that Einstein arrived at the body of thrilling results of his 1905d paper in a 'sudden burst of creativity' and *only after* he had perfected his first three works in the magic spring of 1905. *The basic insight* – the discovery of the relativity of simultaneity – occurred to Einstein only in late May 1905 and only *after* accomplishing the conspicuous 1905c Brownian motion paper. For instance, when many a time and oft inquired by punctilious biographer Carl Seelig, Einstein confessed that:

"Between the conception of the idea of the special theory of relativity and the completion of the corresponding published paper there passed *five or six weeks*" (Seelig 1960, p.114; my italics).

Was it probable that Einstein had relinquished the ether concept on finding out some immaculate, peremptory physical arguments in the writings of those luminaries of science whose influence he publicly admitted? These stout arguments could turn out a final straw for growing aversion to the stale metaphysical remnant of the antediluvian classical tradition.

First, how crucial was Poincaré and Mach's celebrated influence? – For example, in a letter to Michele Besso on 6 March 1951 Einstein recalled:

"These readings [relating to STR] were of considerable influence on my development – along with Poincaré and Mach" (Speziali 1972,

Doc. 182).

Besides, how weighty came about renowned Poincaré's 'Relativity Principle', that alleged relativity of time and space? Already in 1901 Henri Poincaré enounced candourly that

"*There is no absolute time.* To say two durations are equal is an assertion, which has by itself no meaning and which can acquire one only by convention. Not only have we no direct intuition of the equality of two durations, but *we have not even direct intuition of the simultaneity of two events occurring in different places*: this I have explained in an article entitled '*La mesure du temps*'" (Poincaré 1902, p. 114; my italics).

Besides, a permanent member of droll 'Academia Olympia' – Einstein's buddy Maurice Solovine – aptly regarded Henri Poincaré's superlative book "*La science et l'hypothese*" (first published in 1902) as one

"that profoundly impressed us and kept us breathless for many weeks" (Solovine 1956; quoted from Howard and Stachel 2000, p. 6).

Nevertheless, the relativity principle, immaculately molded by Henri Poincaré, *did not debar the latter to believe in the luminiferous ether* as in the medium indispensable for the propagation of electromagnetic waves (Darrigol 2001).

As for Ernst Mach's potent hold, in a letter of 8 April 1951 to biographer Carl Seelig Einstein sincerely admitted:

"My attention was drawn to Ernst Mach's '*Science of Mechanics*' by my friend Besso while a student, *around the year 1897*. The book exerted a *deep* and *persisting* impression upon me owing to its *physical* orientation toward fundamental concepts and fundamental laws" (quoted from Holton 1968, p.636; my italics).

The peremptory influence of astute critique (Mach [1893], 1999, p.229) of Newton's stale concepts of absolute space and time on young Einstein is a humdrum (see, for instance, Frank 1949, p.272). Yet, there is *no* direct and probative way from sublime though the apt philosophical critique of Newtonian mechanics to distinctive STR basic postulates. In my humble judgment, the stoutest argument against the notorious inductivist explanation of the STR origin consists in the next circumstance. Let us recall deft 'emission theories of light' that contested for obvious reasons the inextricable light-constancy postulate and resolutely exchanged it with common Galilean law (that simply added the velocities of light and its source). These artful theories (see for instance Tolman 1912) had *no* problems in comprehending the results of the notorious Michelson-Morley experiment since they were specially concocted to explain it away. And certainly they did. However, they should *not*, if the inductivists were right. Thus, Einstein punctiliously perused Mach's "*Science of Mechanics*" already in 1897;yet it did not hinder him to believe in the luminiferous ether up to 1905.

On the other hand, maybe it was spellbinding David Hume? In a downright letter to Michele Besso in 1948, Einstein confided that

"How far [Mach's writings] influenced my own work is, to be honest, not clear to me, so far as I can be aware, the *immediate* influence of D. Hume on me was great. I read him with Konrad Habicht and Solovine in Bern (quoted from Speziali 1972, p. 153; my italics).

Yet Hume's and Einstein's shrewd conceptions of space and time *differ substantially* (Slavov 2016). In Hume's alluring epistemic doctrine, space and time are *direct* abstractions from simple perceptions. On the other hand, Einstein incessantly maintained that the pivotal concepts of science are *free creations* of the human mind (see, for example, Schilpp's 1949 ample volume and the references cited therein).

3.3 Part two. What was the Salutary Train of Thought that Impelled Einstein to Light Quanta and Whereupon to Special Relativity?

Needless to say, that to render a sound answer one has to dip first into the STR paper itself. The paper (Einstein 1905d) eminently commences with punctuating a profound fissure, a 'deep *asymmetry*' in the theoretical reproduction of electromagnetic induction renowned phenomena. Experience manifests that the induction current engendered in the conductor by the motion of the magnet depends merely on their relative motion. Rather, the Maxwell-Lorentz theory provides two *substantially diverse* accounts of the effect that mysteriously led to *the same* quantitative result. What stands behind this?

Withal, Albert Einstein was *by no means the first* to distinguish profound asymmetries in the theoretical description of the induction phenomenon. In 1885, the asymmetries were noticed by Oliver Heaviside, in 1894 – by Herman Föppl, and in 1898 – by Wilhelm Wien himself (Darrigol 2001, p.377). Therefore, the substantial question is not how Einstein became aware of the asymmetries, but what made them so exasperatingly *intolerable* to him.Einstein sedulously followed Hertz, Heaviside, Wien et al. in recognition that 'something was rotten' in the Maxwell-Lorentz kingdom. Yet he had to furnish a rather distinctive 'diagnosis' and to proffer a jolly peculiar 'cure'.

In my judgment, the conspicuous key to answer the thrilling question lies in *other* works of Albert Einstein and first of all in his stupendous 1905 'annus mirabilis' writings. It is well known that Einstein published *nothing* on the topic of optics and electrodynamics of moving bodies before the wonder year of 1905. More importantly, it was Albert Einstein himself who had just detected another *asymmetry – and of a more profound kind* – in foolhardy 1905a paper '*On a heuristical point of view concerning the processes of emission and transformation of light*'. The paper was published in the same journal '*Annalen der Physik*' yet three months *before* the STR paper. Recall the outset of his 1905a groundbreaking *ouevre*:

"There exists a *profound formal difference* between the theoretical conceptions physicists have formed about gases and other ponderable bodies and Maxwell's theory of electromagnetic processes in so-called empty space" (Einstein 1905a, p.86, my italics).

Nevertheless, to fathom how Einstein arrived at the pivotal cross-contradiction between classical mechanics and electrodynamics, one has to precipitate himself to Max Planck's stupendous efforts first.

As it was already punctuated, at the end of the XIX-th century, three mellow 'paradigms' were doomed to interact strongly in Max Planck's oeuvre: thermodynamics, electromagnetic theory, and statistical mechanics. Incidentally, for him, the role of electrodynamics and statistical mechanics initially was purely instrumental. Maxwell's dainty equations and Boltzmann's intricate statistical technique rendered merely conceptual tools to solve the problem of irreversibility first and only then to grapple with the blackbody radiation inextricable data.

Whatever his original motives, the key, crucial component of Planck's discovery consisted in successive consideration of the electromagnetic radiation through the conceptual lenses of statistical mechanics and thermodynamics. This hallmark of the origin of early quantum theory was punctuated by a Russian prominent theorist of the XX-th century beginning (Goldhammer, 1911 – see the epigraph to chapter II).

Eventually, the ultimate elimination of the cross-contradiction resulted in the contrivance of quantum electrodynamics, in the invention of the quantum theory of radiation that took electromagnetic corpuscles as Boltzmann's gas molecules that can gain energy under collisions with usual molecules and resonators. And the second scholar to notice and advance Planck's pioneering efforts turned out to be Albert Einstein, since he, just as Planck did, was impelled to consider the electromagnetic field as a statistical-mechanical system with an infinite number of degrees of freedom in the process of implementation of his distinctive program of statistical thermodynamics. Thereby he was one of the first to quote Planck's quantum results.

In the first part of the 1905a paper, Einstein shrewdly excavated that the *joint* application of mechanical and electrodynamic Weltanschauungen ('theoretical pictures') for inquiring into the blackbody radiation leads not only to the blunt contradictions with experiments: his paper did not even cite Lummer & Pringsheim or Rubens & Kurlbaum impeccable advances. Moreover, it leads to the staggering *paradox* that cannot be circumvented by common means. To manifest it, Einstein deftly constructs the gedankenexperiment with the abstract objects of both mellow theories. He contemplates an imaginary cavity containing free electromagnetic field, gas molecules, and Planck's fictional resonators and arrives at an ineradicable statement that the joint application of mechanics and electrodynamics *relentlessly leads* to Rayleigh-Jeans law for the energy density of the blackbody radiation. However, "this relation [the notorious Rayleigh-Jeans law] which we found as the condition for dynamic equilibrium does not only lack agreement with experiment, but it also shows that in our picture there can be no question of a definite distribution of energy between Aether and matter", since "the greater we choose the range of frequencies of resonators, the greater becomes the radiation energy in space and in the limit, we get

$$\int\limits_{0}^{\infty} \rho_\nu d\nu = (R/N)(8\pi/L^3)T \int\limits_{0}^{\infty} \nu^2 \rho_\nu d\nu = \infty."$$

(Here R denotes the universal gas constant, N the number of 'real molecules' in one gram-equivalent, T the absolute temperature, L the velocity of light, ν the frequency, and $\rho_\nu \, d\nu$ the energy per unit volume of that part of the radiation whose frequency lies between ν and $\nu+d\nu$).

Although it is commonly held that in the foolhardy paper Einstein was bothered with an explanation of the photoelectric effect, the punctilious study of the startling document elucidates that this evidently was not the case. The measurements of the effect at that time were not sufficiently accurate to point definitely to a clear violation of classical behavior (see Ter Haar 1967 for important details). Whereupon Einstein was bothered not so much by the evidence relating to photoeffect and ardently appealed to fluorescence, photoelectricity, and photoionization data only as to *indirect* evidence in favor of his bold model. On the opposite, Einstein had mostly grappled with the fathoming of the sheer *contradiction* between mechanics and electrodynamics, as well as with the efficacious ways out of it.

So, *what was a judicious reason for Einstein's deep interest in the profound contradictions* between the mellow classical physical theories?

It goes without saying that to render a sober answer one has to delve into the eminent 1946 'Autobiographical Notes' once more since namely this source represents the most systematic recital of Einstein's creative evolution furnished by Einstein himself.

"It was Ernst Mach who, in his *History of Mechanics*, shook this dogmatic faith [in Newtonian mechanics]; this book exercised a *profound influence* upon me in this regard while I was a student. I see

Mach's greatness in his *incorruptible skepticism* and independence; *in my younger years, however, Mach's epistemological position also influenced me greatly...*"(Einstein 1949a, p. 21; my italics).

Mach's "*Science of Mechanics*" is teeming with stubborn attacks on the dominating role of classical mechanics in physics of the day (Mach [1893], 1999, pp. 495, 517). On Mach's strong shoulders, Einstein could therefore freely juxtapose Newtonian mechanics, Maxwellian electrodynamics, and statistical thermodynamics without reducing one to the others. Yet the top-notch component of Machian epistemology (that was lacking in Planck's creativity) but that persistently accompanied Einstein all his scientific career was Mach's distinctive Principle *of Economy of Thought*: 'all science stubbornly aims to replace experience with the shortest possible intellectual operations'. That is why science felicitously condenses an infinity of facts into a single fine law, and why it contrives an extremely concentrated fusion of a multitude of laws in what it calls a theory.

"It is a matter of arranging in systematic order the facts presented that have to be reconstructed by thought to form a system out of them so that each fact may be recovered and reestablished with the least intellectual expense"(Mach [1897], 1984).

Thereafter, in a tardy review of the STR advancement, published in "*Science*" in 1940, Einstein alleges that

"The theory of relativity arose out of efforts to improve, with reference to *logical economy*, the foundation of physics as it existed at the turn of the century"(Einstein 1940/1954, p. 329; my italics. See also Einstein [1933], 1954, p.277; Einstein [1936], 1954, p. 293 and Einstein [1944], 1954, p. 23).

Einstein's true reasons for writing his 1905a paper and its inextricable links with the other 1905 ones can be excavated in his "*Autobiographical Notes*" again. Due to Einstein, the primary stage of 'the revolution begun by the introduction of the field' (Einstein 1949a, p. 37) consisted in the invention of the Maxwellian electrodynamics. All the current pre-Maxwellian studies of physical interactions (the deft theories of Newton, Ampére, Weber, Riemann, et al.) were downright accounts of interactions between several *material points*. Owing to Michael Faraday and James Maxwell, the *Electromagnetic Field* was discreetly inserted into the texture of the XIX-th century physics as a valid component of physical reality equal in rights with the *Material Point*. The hallmark of the problem situation consisted in incisive

"dualism which lies in the fact that the material point in Newton's sense and the field as continuum are used as elementary concepts *side by side*. Kinetic energy and field-energy appear as *essentially* different things" (ibid, p.37; my italics).

Whereupon, just as an inevitable corollary of the dualism,

a "*fundamental crisis* set in, the seriousness of which was suddenly recognized due to Max Planck's investigations into heat radiation (1900).The history of this event is all the more remarkable because, at least in its first phase, *it was not in any way influenced by any surprising discoveries of an experimental nature*"(ibid, p.37; my italics).

Max Planck's alluring result ($E = h\nu$) contradicted the mechanical and electrodynamical basis upon which its derivation depended. Note that according to Einstein,

"My own interest in those years was *less concerned with the detailed consequences* of Planck's results, however important these might be. My major interest was: *What general conclusions can be drawn from the radiation formula* ... concerning the structure of radiation *and even more concerning the electro-magnetic foundations of physics?*" (Einstein 1949a, p.47; my italics).

Thus, Einstein's throbbing attraction in the 1905a paper to the intricate subject of the theory of quanta was provoked by its reconciliation possibilities, by its 'foundation' capacities to arrive at a successful *fusion* of dainty Maxwellian electrodynamics and Boltzmann's intricate statistical thermodynamics. Herewith Einstein was clearly aware that his quantum hypothesis was too foolhardy to be taken literally. Next, it will be demonstrated that he did not maintain "light quanta of energy $E = h\nu$ exist". His formulation was discrete: it would be possible to expose monochromatic radiation in certain limits *as if* it were composed of mutually independent energy quanta.

3.4 Part three. Einstein, Kant, Mach, and a Modicum of Distinctive Duhem.

I reckon that to fathom Einstein's true attitude to light quanta hypothesis in the context of its reconciliation capacities one should commence with the following instructive quotation from Jeroen van Dongen's bona fide book *"Einstein's Unification"*:

"On a number of occasions Einstein expressed himself quite appreciative of Kant's ideas, and some aspects of Einstein's thought did rather *resemble* the Kantian philosophy. Both for instance emphasized the virtue of *striving for unity in science*" (van Dongen 2010, p.49; my italics).

And, in my humble opinion, for *basic milestones* of Einstein's 1905 creativity Kant's influence should be taken as ineradicably crucial for the following reasons.

Firstly, the very possibility of Kant's substantial influence on young Einstein is obvious: Kantian philosophy was extremely influential among the educated classes in Germany in the late XIX-th century, and the extent to which it was professed in German high school was even 'overwhelming' (Beller 2000, p.84). Whereupon, the Neukantianismus (neo-Kantianism with its blunt shibboleth "Zurück to Kant!") was the dominating philosophical movement in German high school from the 1870s until the First World War. Einstein first read Kant at the age of thirteen and again at the age of sixteen (see Howard 1994, p.49 for expressive details).

Thereafter, being an Eidgenössiche Technische Hochshule (ETH) student in Zurich, he had an opportunity to continue his Studium von Kant, Kant's superlative Nachlaß in the winter semester of 1897 at the lecture course on the 'Theory of Scientific Thought' of assistant professor Dr. August Stadler, a prominent neo-Kantian of Marburg school (Einstein 1987, pp. 45-50). The Marburg school was distinguished by its keen interest in foundational and methodological aspects of current science within the peremptory Kantian framework. Assistant professor Dr. August Stadler was Herman Cohen's first doctoral student at Marburg, and Einstein elected to enroll also in Stadler's general course on Kant.

Hence, it is not accidental that a majority of outstanding German physicists hinging upon the 'electrodynamics of moving bodies' hold Kantian or neo-Kantian philosophical standpoints. For instance, the methodological principles for classical mechanics rebuilding crucial for late Heinrich Hertz were to be found by him in Kantian epistemology (Hertz 1899, 1,p.23). Note that even before he met such a convinced Neo-Kantian teacher as Hermann von Helmholtz, Hertz had attended in Dresden a course on Kantian philosophy.

Hertz's Kantian background manifested itself not only in the epistemological scheme diligently described in "*Principles of Mechanics*". According to Jed Z. Buchwald, already in 1884 Hertz had proposed a version of Maxwell's equations that was free of the dubious Naturwissenschaft ether notion completely (Buchwald 1998, p.278).

And, what is more important, quite unlikely Maxwellian field theory, in Hertz's theoretical scheme the source continued to exist as an entity' in and of itself'. In Hertz's diagram, the material object remains unknown, whereas the inferred field is a well-known one. This diagrammatic inversion encapsulates the originality of Hertz's physics. It was since Hertz ignored the physical character of the object that produced his radiation (Buchwald 1998, p.272). Being a faithful pupil of Helmholtz, Hertz learned to watch for novel interactions between laboratory objects without worrying overmuch about the hidden processes that account for the object's effect-producing power.

Thus the inextricable nature of electromagnetic waves appeared to Hertz as a kind of 'Ding an sich' admitting a variety of interpretations. A researcher discreetly chooses the version that is the simplest one to work with. The most substantial thing is the equations sedulously depicting the relations between the objects under investigation (Hertz 1893, p.21). Hertz's juncture strongly suggests that an evident source of the unification strategy of European boffins of the second half of the XIX-th century surely was Kantian philosophy. It is no wonder that the papers of all-influential for Einstein luminaries were goaded by Neo-Kantian epistemology.

Just to quote the *Introduction* to Poincaré's extremely influential book:

"The aim of science is not *things themselves*, as the dogmatists in their simplicity imagine, but the *relations between things;* outside those relations, there is no reality knowable" (Poincaré [1902], 1905, p. XIX; my italics).

Or, in the same vein:

"The object of mathematical theories is *not to reveal to us the real nature of things;* that would be an unreasonable claim. Their only object is to co-ordinate the physical laws with which physical experiment makes us acquainted, the enunciation of which, without the aid of mathematics, we should be unable to effect" (Poincaré [1902], 1905, 235; my italics).

Forsooth, Ernst Mach's stout empiricism sprung out of a genuinely Darwinian belief that human knowledge is an indispensable product of biological evolution. It was a sheer experience to which early organisms had responded, and it was out of such experiences that the first images of the world were eventually constructed. Ultimately, these constructions became *a priori*, allowing new and more subtle understandings. Furthermore, in a sheer Kantian vein, Mach insisted that the human eye had a mind of its own; one perceives not direct stimuli but *relations* of stimuli. Thus, we do not perceive 'reality' itself but rather conceive the after-effects of our nervous system's adaptation to new stimuli. Antediluvian 'representation' theories of perception, doggedly showing off direct correspondences between appearance and reality, were ineradicably untenable for Mach.

Whereupon, even Mach's crude empiricism drew a lot upon Kantianism. Mach credited his philosophical awakening to fathoming, at age 15, his father's copy of Kant's *"Prolegomena"*. As he confessed later,

"the book made at the time a powerful and ineffaceable impression upon me, the like of which I never afterward experienced in any of my philosophical reading" (Mach [1897], 1984, p.30).

Moreover, Immanuel Kant for a long time remained a single philosopher cited by Max Planck. For instance, in an article with a meaningful title "The *Unity of the Physical Universe*" Planck asks:

"Is there any recognizable difference between their [Copernicus, Kepler, Newton, Huygens, Faraday] 'world' and our 'world picture of the future'? Certainly not. For it is known – after Immanuel Kant –

114

that there is no method of proving the existence of such a difference"
(Planck 1960, p.25).

For obvious reasons, later the degree of Einstein's acquaintance
with the superlative writings of Kant only increased. For example, in
1918 Einstein confidentially reported to Max Born:

"I am reading Kant's *Prolegomena* here, among other things, and
am beginning to comprehend the enormous suggestive power that emanated from the fellow and still does" (quoted from Born 1971, pp.
25-26).

In his 1936 profound *"Physik und Realitat"*, he maintained that

"One may say 'the eternal mystery of the world is its comprehensibility'. It is one of the greatest realizations of Immanuel Kant that the
postulation of a real external world would be senseless without this
comprehensibility" (quoted from Einstein 1954, p. 292).

Likewise, in 1949, famously brooding on the basic principles of
reasoning in theoretical physics, Einstein avowed that

"the theoretical attitude here advocated is distinct from that of
Kant *only* by the fact that we do not conceive of the categories as un-
alterable... They appear to be a priori only insofar as thinking without the positing of the categories and concepts, in general, would be
as impossible as breathing in the vacuum" (Einstein 1949, p. 674; my
italics).

Yet what could attract 1905 Einstein in abstract and murky waters
of Kantian epistemology?

According to the author of the *'Critique of Pure Reason'*, it is
Man's freedom from the World that makes science possible. The sen-
sible world conforms to certain basic laws because the human mind artfully *constructs* it according to certain laws. Constructivist foundation
for scientific knowledge peremptorily implies that a priori knowledge of
' things in themselves' is impossible. Yet, in the stupendous *Appendix
to the "Dialectic" of the first Critique*, Kant longed to furnish a faint
rehabilitation of the ideas of traditional metaphysics (Wolff, Locke) by
professing that the ideas of reason play a lead in the conduct of natural
science. Surely, this is the case, if they are comprehended regulatively,
i.e. if they are taken to incarnate not metaphysical beings or entities,
but rather ultimate *goals and directions* of scientific inquiry.

In 'Critique' Kant resolutely divided the human intellect into three independent and clear-cut faculties of sensibility, understanding, and reason. Whereupon, the pivotal principles governing these faculties belong either to *constitutive* strict principles or vague *regulative* ones. The constitutive principles were elicited as solid rules for the construction of the phenomenal world (e.g. Newton's laws, Maxwell's equations, etc.). Thus, a fine distinction between the constitutive and regulative principles turns out to be drawn in Kant's ne plus ultra book along the following strict lines. The constitutive principles are those stout rules that govern the function of understanding and are necessary conditions of experience, whereas regulative principles – the vague 'ideas of reason' – exert the function of reason and are not instantiated in experience in the same way. Later none other than Hans Reichenbach regarded constitutive principles as *synthetic a priori* ones and alleged that to retain a role for such principles they should be *relativized* and comprehended as a priori only from the perspective of a particular mellow theory like Newtonian mechanics or Maxwellian electrodynamics. Though constitutive principles make up necessary conditions for the possibility of human experience, they do not necessitate our own experience of the world. Hence, these principles define the space of physical possibilities merely: there is no possibility of our having an experience that directly contradicts the constitutive principles. Whereupon, the constitutive approach alone cannot elucidate how it can be rational to withdraw an entrenched conceptual framework (a 'paradigm', a 'research program', a 'research tradition', and so on) in favor of a brand new one (Everett 2015).

Thus, Kant alleged that the sublime Ideas of Reason could play only a *regulative* rather than a constitutive role. That is why they can be applied only heuristically, as an indispensable guide for our investigations, but not *substantially,* as the authentic inner principle of what one finds out. Reason incessantly demands the systematization of our knowledge; it *spurs for unity.* As science advances, it tentatively replaces a 'narrower aspect of experience by a broader' one. 'Yet in experimental physics, – contends Kant, - even the principles according to which we perform experiments must themselves always be derived from the knowledge of nature, and hence from the theory'

116

(quoted from Buchdahl 1969, p.510, n.1). Thereafter, due to renowned theory-laidenness of observations, science sedulously advances by theory unification. As an eminent neo-Kantian of Marburg school had underscored, 'true unity is never thought in things as such, but in intellectual constructions' (Cassirer [1921], 1923, p.36). In genuinely Kantian wake he punctuated that unification was a purely *regulative* demand.

Forsooth, Kant himself professed in the *Appendix to the Dialectic* that science must adopt certain ideas of reason as a *heuristic* ("as if") devices to encourage systematic unity (Kant [1787], 1998, p. 659).Along these lines, Fölsing (1997) shrewdly noticed that Einstein probably first learned to think in terms of his 1905a 'heuristic viewpoint' from a diligent reading of Kant. Einstein's wayward 'heuristic method' was to proffer such a forethought assertion from which familiar facts could then be naturally deduced. Einstein's path-breaking 1905a paper was discreetly entitled '*Über einen die Erzeugung und Verwandlung des Lichtes betreffenden heuristischen Lesichtspunkt*' ('On a *Heuristic Point of View* Concerning the Production and Transformation of Light').

While incessantly dissociating himself from Kantian 'synthetic a priori', Einstein keenly proceeds in the general neo- Kantian epistemic spirit:

"The following, however, appears to me to be correct in Kant's statement of the problem: in thinking, we use, with a certain 'right', concepts to which there is no access from the materials of sensory experiences" (Einstein 1944, p.22).

To sum up, I reckon that the felicitous notion for comprehending important facets of Einstein's 1905 research activity is *Kant's concept of systematic Unity of Nature* as an indispensable regulative idea. This unity, both for Kant and Einstein, is not an ontological tenet at all. It is meaningless to inquire into whether Mother Nature possesses such a unity or not. On the contrary, the idea of unity has incontestable epistemic value, since systematic unity of nature provides a downright benchmark *of validity for the scientific hypothesis* that complements the empirical idea of confirmation. Not by a chance, Kant ([1787], 1998, p.592) displays the 'hypothetical employment of reason', emphasizing once and oft that the confirmation of a hypothesis by its empirical consequences can never endow such a hypothesis with universality, or 'certainty': 'In natural science... there is endless conjecture, and certainty is not to be counted upon' (Kant [1787], 1998, p.608).

Since a given hypothesis of natural science cannot obtain the sheer proof of its truth from 'below', from repeated experimental confirmation, something else is necessary. One needs the efficacious *criterion* that can distinguish contingent and unimportant empirical generalizations from genuine profound Laws of Nature, which are endowed with Universality and Necessity. The Kantian fruitful idea of the 'truth' of a proposition is equivalent to its being a law-like statement.

"Such concepts of reason are not created by nature, rather we question nature according to be deflective as long as it is not adequate to them [...] The hypothetical use of reason is therefore directed at the systematic unity of the understanding's cognition, which, however, is the *touchstone of truth for its rules*" (Kant [1787], 1998, p. 592; my italics).

Hence, from the hotchpotch of multifarious uniformities, only those can be taken as having a law-like necessity that can be fitted into a unified, *systematized* general system. Accordingly, in the '*Autobiographical Notes*', brooding on his methodological principles, Einstein stressed that

"A system has truth-content according to the certainty and completeness of its coordination-possibility to the *totality of experience*. A correct proposition borrows its 'truth' from the truth-content of a system to which it belongs" (Einstein 1946, p. 13; my italics).

Incidentally, the vague term 'totality of experience' seems to point on the other significant source of Einstein's philosophical inspiration - Duhem's distinctive conventionalism.

"In sum, the physicist can never subject an isolated hypothesis to experimental test, but only a *whole group of hypotheses* [...]. Physical science is a system that must be taken as a *whole*; it is an organism in which one part cannot be made to function except when the parts that are most remote from it are called into play" (Duhem [1905], 1954, p.187, my italics; see also Quine [1951], 2000, p.345).

Note that it was this 'holistic' farsighted stand that allowed Einstein as early as in 1906 to call into question the widely accepted results of notorious Kaufmann's 'crucial experiments'. These data hastily seemed to buttress the Abraham-Bucherer deft theory and to refute the 'Lorentz-Einstein' one (Holton 1968, p.253; Miller 1981, p.124).

As Einstein had perspicuously put it, the rival theories (e.g. Abraham's electron theory)

"have rather small probabilities, because their fundamental assumptions (concerning the mass of moving electrons) are not explainable in terms of theoretical systems which embrace a greater complex of phenomena" (Einstein as quoted and translated in Holton 1968, p. 253).

Einstein consented that empirically equivalent alternative theories can be invented for any realm of natural phenomena. In this respect, his views were close to Pierre Duhem. The latter contended:

"the validity of a theory is measured by the number of experimental laws it represents and by the degree of precision with which it represents them; if two different theories represent the same facts with the same degree of approximation, the physical method considers them as having the same validity; it does not give the right to dictate our choice between these two equivalent theories and is bound to leave us free. Grounds of suitability which are essentially *subjective*, contingent, and variable with time, with schools and with" (Duhem 1906, 1954, p. 288).

Yet if for such a confounded conventionalist as Pierre Duhem this ' underdetermination of theory by facts' meant that there can be no ultimately true theory, the Einsteinian train of thought in the Kantian wake not only rendered a meaning to the regulative ideal of a final theory. It also highlighted Einstein's significant remarks that despite this exasperating underdetermination at any given time there is *only one* true theory: the theory with the *greatest power of unification* (Einstein 1918, p. 226).

Thus, Einstein's full involvement in the 1905a paper to the intricate subject of the theory of quanta was engendered first by its peremptory unifying *possibilities*, by its fascinating capacities to buttress efficacious fusion of Maxwellian electrodynamics and Boltzmann's statistical thermodynamics. Hence in the ground-breaking 1905a, he starts with the heart of what troubled him most – the *Deep Abyss* in the foundations of physics felt most incisively in Lorentz's Electron Theory (and 'H.A. Lorentz [himself] knew this very well'; Einstein 1949a, p.37).

3.5 Part four. A Bed of Thorns: The Distinctive Way From Light Quanta to Special Relativity.

To sum up: h*ow had 1905 Einstein intended to get rid of the pivotal contradiction of his 1905a paper?*

While brooding over Einstein's way out of the obstacle, one should take into consideration that *all* Einstein's papers from 1901 to 1905 possess one hallmark in common: the statistical-thermodynamics approach. Thomas S. Kuhn even professed that what impelled Einstein to the idea of the photon *was an incessant development of a research program started in 1902*, a program 'so nearly independent of Planck that it would almost certainly have led to the black-body law even if Planck had never lived' (Kuhn 1978, p. 171).

Forsooth, from the outset of his dizzying scientific career, Einstein was fascinated by the simplicity and scope of classical thermodynamics. Yet for him thermodynamics embraced the statistical approach he had imbibed from Boltzmann's aspiring writings, and so he strived to unfold statistical thermodynamics. The promising result was a series of three papers published in 1902, 1903, and 1904. Expressly *they (*see especially Einstein 1902, 1903) provide *a solid clue* for comprehending his 1905a paper on quanta, his 1905b dissertation, 1905c work on Brownian motion, and 1905d paper on STR. Just to quote the first of Einstein's three papers on the foundations of statistical mechanics submitted to *'Annalen der Physik'* in June 1901 where his ambitious aim is boldly enunciated.

"Great as the achievements of the kinetic theory of heat have been in the domain of gas theory, the science of mechanics has not yet been able to produce *an adequate foundation for the general theory of heat,* for one has not yet succeeded in deriving the laws of thermal equilibrium and the second law of thermodynamics using only the equations of mechanics and the probability calculus, though *Maxwell's and Boltzmann's theories* came close to this goal. The purpose of the following considerations is *to close this gap*" (CPAE, vol.2, Doc 3, 57; my italics).

Hence, the fact that Einstein took up the blackbody problem was not accidental. He would still come to the problem because of the implementation of his own program 'even if Planck had never lived'. Besides, the first substantial result consisted in that for physical systems of jolly general sort Einstein had rendered, by the summer of 1903, both a generalized measure for temperature T and entropy S, embracing some universal constant χ. By the time he accomplished his 1903 paper, Einstein had fathomed that χ could be evaluated in terms of the values of the gas constant and Avogadro's number. Yet the theory that had incited him to invoke the constant was, however, applicable to systems far more general than gases. It should, therefore, possess a correspondingly general physical foundation, engendered by a statistical-mechanical approach that led him to the constant, properly explaining not only its lead as a scale factor for temperature but also its position as a multiplier in the probabilistic definition of entropy. The physical significance of χ turned out to be the pivotal problem regarded in Einstein's third statistical paper "*On the General Molecular Theory of Heat*", submitted to the '*Annalen der Physik*' in the spring of 1904. The solution to the problem was inextricably connected with the queer phenomena of energy fluctuations. Einstein expounded that

$$\bar{\varepsilon}^2 = 2\chi T dE/dT,$$

where $\bar{\varepsilon}$ is a measure of the thermal stability of the system, T – the temperature of the system, and E its energy. Moreover, it was conceiving the constant physical sense that directed his attention to the blackbody problem.

"The equation just found would permit an exact determination of the universal constant χ if it were possible to determine the energy fluctuation of the system. In the present state of our knowledge, however, that is not the case. Indeed, *for only one sort* of physical system can we presume from experience that an energy fluctuation occurs. That system is empty space filled with thermal radiation" (Einstein 1904, p.360, my italics; translated in Kuhn 1978).

At least one more step in the distinctive program of statistical thermodynamics was needed, and Einstein deliberately took it in the consequent 1905a paper. Its content makes one think that Einstein had commenced seeking a blackbody law of his own, and that, due to Planck's writings, he had elicited the paradox, evinced in the cross-contradiction between statistical mechanics and Maxwellian electrodynamics. Therefore, he had dropped the search for the law in favor of proper exploration of the thrilling paradox itself. It is clear from the outset of already quoted paper (excellently translated in Ter Haar 1967). The first part of the 1905a ended by showing the 'ultraviolet catastrophe' off. *Yet how did Einstein intended to salve the paradox?* To give a definite and sober answer, one is impelled to return to the ideas of the patriarch of the German school of theoretical physics, the editor of "Annalen der Physik", who scrutinized, edited, and recommended for publication both Einstein's 1905a and 1905d papers. The key lies in the following words that indicate that the problem of the gap between the profound classical traditions and the ways of its elimination was at the center of his attention:

"In physics as a practical science, it has frequently happened and still does happen that two theories, developed independently of one another, *come into conflict* when extended and *must be mutually modified to remain compatible.* In this mutual adjustment lies the germ of their further development into complete unity" (Planck 1960, p.83; my italics).

The key instruction for the elimination of the cross-contradiction is contained at the beginning – '*must be modified to remain compatible*'. Note that Planck considers mechanics and electrodynamics not as incommensurable but as similar in structure and content. Therefore, one should not destroy them; one has to *modify* them only. Viz.,

"The numerous independent domains of the earlier physics now appear reduced to two: mechanics and electrodynamics, or, as one may say: *the physics of material bodies* and *the physics of the ether*. The former comprehends acoustics, phenomena in material bodies, and chemical phenomena; the latter, magnetism, optics, and radiant heat. But *is this division a fundamental one?"* (Planck [1910], 2015, p.4; my italics).

Planck is emphatically optimistic:

"For myself, I believe it must be answered in the *negative,* and upon the following grounds: *mechanics and electrodynamics cannot be permanently sharply differentiated from each other.* Does the process of light emission, for example, belong to mechanics or electrodynamics? To which domain shall be assigned the laws of motion of electrons? At first glance, one may perhaps say: to electrodynamics, since with the electrons ponderable matter does not play any role. But let one direct his attention to the motion of free electrons in metals. There he will find, in the study of the classical researches of H.A. Lorentz, for example, that the laws obeyed by the electrons belong to the kinetic theory of gases than to electrodynamics. In general, it appears to me that *the original differences between processes in the ether and processes in material bodies are to be considered as disappearing. Electrodynamics and mechanics are not so remarkably far apart,* as is considered to be the case by many people, who already speak of a conflict between the mechanical and electrodynamic views of the world" (Planck [1910], 2015, p.4; my italics).

On Planck's experienced, sophisticated view, Newtonian mechanics and Maxwellian electrodynamics are not *incommensurable*: mechanics is grounded on the ideas of space and time, yet the same can be told on the classical electrodynamics. Hence

"In fact, there are many indications pointing toward the *ultimate amalgamation* of these two subjects, the domains of which already overlap in some measure" (Planck [1910], 2015, p.4; my italics).

Consequently, a man can enter the scene who strives to provide the situation when 'the gulf between ether and matter will be once bridged' (Planck [1910], 2015, p.4). And the man was Albert Einstein. According to his *"Autobiographical Notes"*,

"Reflections of *this type* [i.e. on the molecular structure of radiation] made it clear to me as long as shortly after 1900, i.e. *shortly after* Planck's trailblazing work, that *neither mechanics nor electrodynamics could (except in limiting cases) claim exact validity"* (Einstein 1949a, p.51; my italics)

Hence classical mechanics and electrodynamics are both ultimately wrong, being only limiting cases of a general theory, and a way out for constructing the global theory – to change mechanics to fit electrodynamics and to change electrodynamics to meet mechanics.

Besides, to demonstrate it in the subsequent part of his 1905a paper, Einstein sedulously applies thermodynamics, statistical mechanics, and Maxwellian electrodynamics to examine the realm of empirical reality covered by Wien's radiation law. He takes $\beta = h/k = Nh/R$ as an undefined constant in the 1905a paper and hence he writes $R\beta/N$ everywhere instead of h. (R denotes the universal gas constant, N the number of 'real molecules' in one gram-equivalent, h is Planck's constant and k is Boltzmann's one). The downward application of the three mellow theories, belonging to three profound research traditions of classical physics, enables Einstein to draw the following startling conclusion. If monochromatic radiation of frequency ν and energy E is enclosed in the volume V_0, then the probability W that at any moment all the radiation energy will be detected in the partial volume V of the volume V_0 is given by

$$W = (V/V_0)^{E/h\nu} \qquad (i)$$

Yet in the same paper, Einstein had previously made sure that, in the case of n independently moving gas particles enclosed in a volume V_0, the probability of discovering them all momentarily in the sub-volume V is

$$W = (V/V_0)^n \qquad (ii)$$

Comparing eq. (i) and (ii), Einstein draws a startling yet discreet conclusion that '*monochromatic radiation of small density behaves in thermodynamic respects as though it consists of distinct independent energy quanta of magnitude $h\nu$*'.

Thus, the assertion that radiation in the cavity 'consists' of independent energy quanta follows *directly* from the immaculate application of general principles of thermodynamics and statistical mechanics to whimsical radiation phenomena.

However, in the turn of the XIX-th century, all the available experimental data, relevant to fluorescence, photoelectricity, and photoionization, would guarantee only *an indirect* verity of the quantum hypothesis. Hence, to put to the severe test the ultra-revolutionary hypothesis of quanta, Einstein had to appeal to a *'crucial experiment'* of a very peculiar, freaky kind. (See Chapter I where we have punctiliously discussed the juncture in modern relativistic astrophysics, where the entropy of a black hole was derived from the superstring theory). Likewise, Einstein was impelled to compare queer quantum results with the assertions of another mellow 'old' theory invented independently of the 1905a bizarre hypothesis. Incidentally, this theory had to be sufficiently 'old' and worked out to accumulate the results of a heap of experiments. So, if the 1905a results had matched the consequences of the fairly diverse theory that arose out of substantially different problem situations, they would have rendered an especially reliable, solid verification of daring 'photon hypothesis'. Let us recall that

"A proposition is correct if, within a logical system, it is deduced according to the accepted logical rules. A system has truth-content according to the certainty and completeness of its coordination-possibility to the *totality of experience*. A correct proposition borrows its 'truth' from the truth-content of a *system* to which it belongs" [Ein richtiger Satz erborgt seine 'Wahrheit' von dem Wahrheits-Gehalt des Systems, dem er angehört] (Einstein 1949a, p.13; my italics).

In the opposite case, the 1905a theory would have been 'falsified' not by a single 'critical experiment', but by a whole pile (multitude) of the well-established experimental data. Incidentally, this 'holistic' quasi-Duhemian standpoint allowed Einstein, as early as in 1906, to renounce the exasperating results of Kaufmann's 'crucial experiments', which seemed to support deft Abraham-Bucherer theory and to refute the 'Lorentz-Einstein' creation (Holton 1968, p.253; Miller 1981, p.124).

As Einstein recalled later, the rival theories (e.g. Abraham's electron theory)

"Have rather small probabilities, because their fundamental assumptions (concerning the mass of moving electrons) are not explainable in terms of theoretical systems which embrace a greater complex of phenomena" (Einstein as quoted in Holton 1968, p.253).

Thus, the next - 1905b - result was apparently decisive for the 1905a actual verification. In the 1905b paper, Einstein sedulously perpetrated the principles of Brownian motion that were directly verified by immaculate Perrin's experiments.

"My principal aim in this [1905b work on Brownian motion] was to find facts that would guarantee as much as possible the existence of atoms of definite size... The agreement of these considerations with experience together with Planck's determination of the true molecular size from the law of radiation (for high temperatures) convinced the skeptics, who were quite numerous at that time (Ostwald, *Mach*), of the *reality of atoms*" (Einstein 1949a, pp.45-47; my italics).

Though the importance of 1905b paper's for the 1905a one was admitted by Einstein much later; he credulously confided to Max von Laue on 17 January 1952:

"When one goes through your collection of verifications of the special relativity theory, one believes that Maxwell's theory is firmly established. But in 1905 I knew *already with certainty* that it leads to the wrong fluctuations in radiation pressure, and consequently to an incorrect Brownian motion of a mirror in a Planckian radiation cavity" (quoted from Rynasiewicz 2000, p.177; my italics).

This apparent for 1905 Einstein result was posited to the scientific community only in 1909 when Einstein masterly applied his theory of Brownian motion to the case of a two-sided mirror immersed in thermal radiation. He manifested that the mirror would be unable to carry out a Brownian motion indefinitely, if the fluctuations in the radiation pressure on its surfaces were solely due to the effects of random waves, as predicted by Maxwell's theory. Nevertheless, only the presence of an additional term, corresponding to pressure fluctuations due to the impact of random particles, ensures the continued Brownian motion of the mirror. Einstein exhibited that similar fluctuation terms in the energy constituted the definite consequences of Planck's law. He decidedly took such fluctuation phenomena as the *strongest argument* for ascribing physical significance to the queer light quanta (Stachel 2000). Only after this whimsical 'crucial thought experiment', that is, only *after* the 1905b paper, could Einstein look forward to examining the startling consequences of his intricate light quantum hypothesis; and so he finally returned to his half-forgotten 'unsere Arbeit uber die Relativbewegung', eine 'kapitale Abhandlung'. This reconsidered odd manuscript was destined to become the celebrated STR. So far, so good.

"If the monochromatic radiation (of sufficiently small density) in the sense of entropy dependence upon volume behaves itself as a discontinuous medium, consisting of energy quanta $R\beta\nu/N$, a question occurs: if they are not the laws of creation and conversion of light such as if it consists of similar energy quanta?" (Einstein 1905a, p. 236). That is the subject cautiously broached by Einstein at the *end* of §6 of his 1905a paper. However,*the ether conception turned out to be a substantial setback* . It extinguished the positive answer and put *insurmountable obstacles* in implementing Einstein's statistical-thermodynamics program. Forsooth,

"Mechanical and purely electromagnetic interpretations of optical and electromagnetic phenomena have in common that in both cases the electromagnetic field is considered as a special state of the hypothetical medium filling all the space. Namely in that point, two interpretations mentioned *differ radically from Newton's emission theory, in which light consists of moving particles.* According to Newton, space should be considered as possessing neither ponderable matter, nor light rays, i.e. absolutely empty" (Einstein 1905a, p.236; my italics).

To put up a solid quantum theory of radiation, one needs electromagnetic fields as *independent* entities that can be emitted by the source 'just as in Newton's emitting theory'. It means that the energy conveyed in a process of emission should not be dissipated in space, but should be *completely* preserved until an elementary act of absorption. Yet within the Lorentz reductionist program, an electromagnetic field is regarded as a specific state of ether - a state of medium that is *continuously* distributed in space. In such a stout medium, an elementary process of radiation is connected only with a *spherical* wave.

Withal, repugnance to the ether and acceptance of emission theory should lead to Walter Ritz's 1908 sleight-of-hand 'ballistic hypothesis', wherein the velocity of quantum should depend on the velocity of its source. In Ritz's deft theory the velocity of light is not constant but is equal to v+c, where v is a relative velocity of the observer and the source.

Later, in April 1922, Einstein had confided to Viscardini:

"I rejected this [emission] hypothesis at that time because it leads to tremendous theoretical difficulties (e.g. the expectation of shadow formation by a screen that moves relative to the light source)" (quoted from Rynasiewicz 2000, p.182).

Thus Einstein, by contrast, never thought of renouncing from Maxwell's entrenched theory, just as Newton, the inventor of the distinctive emission theory, did not reject the wave theory 300 years earlier. Hence, in 1905a paper Einstein had to make the following reservation:

"Wave theory operating with point continuous functions is *excellently* justified when describing purely optical phenomena and perhaps would not be replaced by another theory" (Einstein 1905a, p.237; my italics).

In Lorentz's alluring theory, this obstacle was absent. Indeed, in the reference frame that is at rest relative to the ether light propagates with constant velocity c independent *of the velocity of the source*. Hence, if one craves to give the idea of ether up, yet to come to terms with Maxwell's well-established theory at the same time, s/he should disown queer ballistic hypothesis. He/she should 'raise to the rank of a [general] principle the validity of the law of constancy of light velocity for all inertial frames' (Einstein [1936], 1954, p.307), i.e. to set up a special *'principle of the constancy of the velocity of light'* (I).

The second celebrated STR postulate - *'the principle of relativity'* (II) - follows immediately from the assertion that there is no luminiferous ether and, consequently, no absolute system of reference.

Just as Einstein famously underscored in the *'Autobiographical Notes'*,

"By and by I despaired of the possibility of discovering the true laws by means of constructive efforts based on known facts. The longer and the more despairingly I tried, the more I came to the conviction that only the discovery of a *universal formal principle* could lead us to assured results. The example I saw before me was *thermodynamics*"

(Einstein 1949a, p.51; my italics).

The latter point needs further clarification via delving into Mach's pre-eminent *"Mechanics"* - the stout source of 1905 Einstein's vision of the history of physics. The most alluring case study of the interconnection between the principle of economy of thought and the second law of thermodynamics in *"Mechanics"* is Stevinus's (1548-1620) solid theoretical scheme of statics. In his immaculate *"Hypomnemata Mathematica"* (Leyden 1605), Stevinus punctiliously examined the mechanical properties of the inclined plane. He strived to set up a stout general theoretical principle and then to inquire into its partial cases that can be easily treated by quantitative means. To furnish the pivotal gedankenexperiment, necessary to elicit his general principle, Stevin conjures up an imaginary triangular prism with no horizontally placed edges. Over the prism, he lays an endless string on which 14 balls of equal weight are strung and tied at equal distances apart. (The string can be readily replaced by an endless uniform chain).

This chain will either reside in equilibrium or not. If one assumes the latter to be the case, the chain must when once actually in motion, continue to move perennially. In other words, it must engender perpetual motion, which Stevin maintains absurd. Hence, only the first case is conceivable, and *the chain always remains in equilibrium.*

Mach stresses that in the main premise from which Stevin starts that *the endless chain does not move,* only a *purely instinctive* cognition is contained. He realizes at once, and the readers with him, that one have never seen anything like the motion of this kind. This conviction has so much logical cogency that one easily accepts the conclusion drawn from it respecting the law of equilibrium on the inclined plane without an objection, *although the law is shrewdly rendered as an evident experimental result.* One cannot be surprised at this when s/he takes into consideration that all results of the experiment are obscured by adventitious circumstances (such as friction, etc.) and that every conjecture as to the conditions, which are determinative in a given case, are liable to error. Amazingly, Stevinus ascribes to the abovementioned instinctive knowledge a *higher authority* than to simple, manifest, and direct observations.

Whereupon one is faced with the following substantial problem:

w*hence does this higher authority come?* If we recall that scientific demonstration and scientific criticism generally can only have sprung from the consciousness of the individual fallibility of researchers, the explanation is evident. We are aware that we have rendered *nothing* to the creation of this distinctive 'Instinctive *Knowledge*', that we have added to it nothing, but that it exists in evident independence of our participation.

According to Ernst Mach, Stevinus's deft deduction is one of the rarest 'fossil indicators' that one possesses in the history of classical mechanics. Whereupon, it throws a thrilling light on the subtle process of the formation of science generally, on its remarkable rise from instinctive knowledge.

Nevertheless, every true experimenter can daily reveal in his person the guidance that Instinctive Knowledge furnishes him. If he succeeds in *abstractly formulating* what is contained in it, he will have made a substantial advance in science. And it is perfectly beyond doubt for the author of "*Mechanics*" that it is the rarest combination of the strongest instinct with the greatest power of abstract formulation alone that makes up the great natural inquirer (Mach [1893], 1999: p.27].

However, how does this intricate 'instinctive knowledge' originate and what is its fine structure?

Everything, which one observes in nature, imprints itself *uncomprehended* and *unanalyzed* in his distinctive percepts and clear ideas. In the accumulated experiences, he possesses a sheer 'treasure store' that is ever close at hand and of which only the insignificant part is embodied in fine articulate thought. The fact that it is far easier to appeal to these experiences than it is to Nature herself, and they are, notwithstanding this, free from all the vague subjectivity, *invests them with a high value.*

"It is a peculiar property of instinctive knowledge that it is predominantly of a *negative* nature" (Mach [1893], 1999: p.28].

One cannot so well profess what must happen as one can what cannot happen. Whereupon the latter alone stands in devastating contrast to the obscure hotchpotch of experience in us in which single characters are not distinguished.

Moreover, contends Mach, another hallmark is no less important for epistemology and philosophy of science. The thrilling reasoning of Stevinus has such a strong influence upon us because the result at which he arrives contains *more* than the premise from which he starts.

Incidentally, it often happens in the course of the advancement of science, that a new principle perceived by some researcher in connection with a fact, is not realized and set up immediately in all its generosity. If throughout all facts, we elicit and *discern* a principle, which, though not admitting of proof, can yet be known to dominate, we have advanced much farther in the consistent view of nature than if we suffered ourselves to be overawed by a specious demonstration (Mach [1893], 1999, p. 82].

In my judgment, *all the above-mentioned Stevin-Mach fascinating recipes* where deftly implemented by Albert Einstein in formulating the basic STR principle – the principle of relativity. However, due to the foolhardy and extremely speculative nature of the light-quanta hypothesis, he could not dare to manifest the indissoluble link with the 1905a paper directly. Hence, he used all the Stevin-Mach artful technique of conviction at hand to lay out his electrodynamics of moving bodies *in sheer phenomenological wake.* One should pay special respect to the *negative character of the relativity principle* and the manner of its connections with experiments and observations that are closer to the subtle conviction technique of instinctive knowledge than to blunt inductive way of inference. Behold once more the renowned beginning of the STR paper:

"Examples of a similar kind, and the failure of attempts to detect a motion of the earth relative to the 'light medium', lead to the *conjecture* that not only in mechanics, but in electrodynamics as well, the phenomena *do not* have any properties corresponding to the concept of absolute rest, but that in all coordinate systems in which the mechanical equations are valid, also the same electrodynamic and optical laws are valid, as have already been shown for quantities of the first order. We shall raise this conjecture (whose content will be called 'the principle of relativity' hereafter) to the status of a postulate and shall introduce, in addition, the postulate, only seemingly incompatible with the former one, that in empty space light is always propagated with a definite velocity V which is independent of the state of motion of emitting body" (Einstein 1905d, p.140; my italics).

By the way, the artful appeal to *instinctive knowledge* easily explains the fact that the STR paper by rights stands out in all the world scientific literature for the *complete lack of references*. According to the "*Autobiographical Notes*", Einstein's novel crossbred theory was contrived because of the inevitable encounter of classical mechanics and Maxwellian electrodynamics. Hence, its basis should consist of (truly 'Duhemian') *minimum* of two postulates, (1) the first drawn from Newtonian mechanics (the principle of relativity) and (2) the second one directly transferred from the Maxwell-Lorentz electrodynamics (the principle of the constancy of the velocity of light). Viz.,

1. "*Classical mechanics*, of which it could not be doubted that it holds with a close degree of approximation, teaches the equivalence of all inertial systems or inertial 'spaces' for the formulation of natural laws, i.e., the invariance of natural laws with respect to the transition from one inertial system to another" (Einstein 1954, p.369).

2. "This [the STR] takes over from the *theory of Maxwell-Lorentz* the assumption of the constancy of the velocity of light" (Einstein [1940], 1954, p. 370).

In my view, to conceive an origin of both postulates, and especially the first one, we may appeal to the recipes of another Mach's contemporary and accomplice:

"We recognize the correct principle by the facility with which it straightens out the complicated difficulties into which the use of erroneous principles brought us. If, therefore, the idea we have put forth is correct, namely, that comparison is established necessarily between the *whole* of theory and the *whole* of experimental facts, we ought in the light of this principle to see the disappearance of the obscurities in which we should be lost by thinking that we are subjecting each isolated theoretical hypothesis to the test of facts" (Duhem [1906], 1954, p.208; my italics).

The two basic postulates, (I) + (II), the relativity principle plus the principle of the constancy of the velocity of light, are sufficient, according to Einstein, to work out the Duhem-Mach amended version of electrodynamics of moving bodies. Yet, since

"The theory based on these two principles should not to lead to contradictory results, one must renounce the customary rule of addition of velocities" (Einstein 1910, p.125).

Namely, that was brought to fruition in the 1905d paper "*On the Electrodynamics of Moving Bodies*", published several months *after* the foolhardy 1905a one. Einstein had masterfully elicited the hidden assumption - the basis of the Galileo addition law - that the statements of time, as well as of the shapes of moving bodies have the sense independent of the state of motion of the reference frame. He persuasively highlighted that the acceptance of the 'principle of relativity' together with the 'principle of the constancy of light' is equivalent to significant modification of the simultaneity concept and to thrilling clock delay in moving reference frame.

Note that, besides Mach and Kant, the ideas of other philosopher were applied to protect the STR postulates from the scarcing criticism of the opponents. First, one could recall Pierre Duhem's eminent contra-inductivist dictum that was quoted already (Duhem 1954, p.19).

Yet a more complete justification of STR postulates may necessitate the application of the following recipes of the author of "*Aim and Structure of a Physical Theory*":

"What are then the conditions logically imposed on the choice of hypotheses to serve as the base of our physical theory? These conditions are three in number.

In the first place, a hypothesis shall not be a self-contradictory proposition, for the physicist does not intend to utter nonsense.

In the second place, the different hypotheses, which are to support physics, shall *not contradict one another*. Physical theory, indeed, is not to be resolved into a mass of disparate and incompatible models [as in Maxwell's theory]; it aims to preserve with jealous care a logical unity, for an intuition we are powerless to justify, but which we can't be blind to, shows us that only on this condition will theory tend towards its ideal form, namely, that of natural classification.

In the third place, hypotheses shall be chosen in such a manner that from them taken as a whole mathematical deduction may draw consequences representing with a sufficient degree of approximation the totality of experimental facts" (Duhem 1954, p.220).

It should be underscored that in *no* way 1905 Einstein was an idle thinker brooding on the slipping off essences of space and time. He was *forced* to inextricable philosophical reasoning on the nature of Space and Time by his common *research practice*, by a mundane physical problem of reconciling classical mechanics (the Principle of Relativity) with classical electrodynamics (the Light Constancy Postulate).

Hence, at least, in that case, Einstein's implementation of Hume and Mach's ambiguous philosophical writings was *'highly selective'*, forsooth (Norton 2010, p.359). His ultimate goals were not so much to apprehend Hume's and Mach's lofty and multifocal reflections as to find in them fruitful ideas that may be useful in his mundane research practice.

Well, if all said before is true, the following blunt question should be answered: why Einstein in the 1905d paper *did not cite his 1905a breakthrough* on light quanta? To furnish a sober, one has to dwell into Einstein's 1905 private correspondence now. Writing to his close friend Conrad Habicht in 1905 and sending him the thrilling results of his labors at that time, Einstein proudly called his light quanta paper *'very revolutionary'*, while the STR one was humbly characterized as *'interesting* in its kinematical part'. So, a reference in the paper, making actual substantial changes yet mainly of a metaphysical character, on the foolhardy hypothesis that had already rendered *revolutionary changes and had blatantly contradicted Maxwell's ingrained theory,* could hardly make the arguments stronger.

An excellent mark of the intricate state of the problem situation is Max Planck's attitude to both 1905a and 1905d papers. As a chief editor of *"Annalen der Physik"*, he had recommended both of them for publication. However, though he tentatively distasted the idea of 'light quanta', simultaneously he was one of the first to approve the STR. Hence Einstein ñcomplained in 1907 to his friend Jacob Laub that

"He [Planck] has, however, one fault: that he is clumsy in finding his way about in foreign trains of thought. It is therefore understandable

when he makes quite faulty objections to my latest work on radiation"
(Seelig 1960, pp.102-103; quoted from Weinstein, 2013).

Planck himself confided to Einstein on July 6, 1907:

"In any case, I do not believe that this difference in our opinion is
of fundamental nature.

3. But things may perhaps be different when it comes to the following question: Does the absolute vacuum (the free ether) possess any
atomistic properties? Judging by your remark (Ann 23, p.372, 1907)
that the electromagnetic state in a portion of space is determined by
a finite number of quantities, you seem to answer the question in the
affirmative, while I would answer it, at least in line with my present
view, in the negative. For I do not seek the meaning of the quantum
of action (light quantum) in the vacuum but at the sites of absorption and emission, and assume that the processes in the vacuum are
described exactly by Maxwell's equations" (Einstein, 1993, Doc.47).

Thereafter, in a 1910 paper on the theory of thermal radiation,
Max Planck eloquently advised:

"It appears to me that utmost caution against the new Einsteinian
corpuscular theory of light would be warranted... The theory of light
would be thrown not decades, but centuries, to the time when Christian Huygens dared to take up his fight against Newton's overpowering
emission theory... And all the accomplishments, which are among the
proudest successes of physics, indeed of scientific research overall, are
supposed to be sacrificed for the sake of some still quite contestable
observations? Heavier artillery would need to be run out to sway this,
by now, *very firmly founded* edifice of electromagnetic light theory"
(Planck 1910; quoted from Hentschel 2018, p.21; my italics).

In my view, the mark of Einstein's fear to disclose the connections between 1905a and 1905d is the fact, that in his STR paper he
prudently exchanged the term 'light quanta' by his newly – coined
euphemism 'light complex' for which no clear definition was given.
Galina Weinstein (2013) draws attention to Einstein's conclusion that

"It is noteworthy that the energy and the frequency of a light complex vary with the observer's state of motion according to the same
law" (Einstein 1905d, p.914).

Accordingly, John Stachel (2008) shrewdly noted that the 1905d formula $E'/E = \nu'/\nu$ corresponds to the light quanta hypothesis and is intended to buttress the latter. On his view, Einstein's ultimate aim was to manifest that the startling equation $E = h\nu$ takes the same form in any inertial frame. Viz., $E=h\nu$ is easily transformed to $E'=h\nu'$ blatantly positing the invariance of the relativity postulate.

Respectively, Robert Rynasiewicz aptly enunciates that "Einstein wraps up his derivation with what is an allusion to the light quantum hypothesis' (Rynasiewicz 2005, p.47).

Einstein himself at the epoch-making first Solvay Congress had to admit publicly (through his teeth)

"provisional character of this concept [light quanta] which does not seem reconcilable with the experimentally verified consequences of the wave theory" (quoted from Pais 1979, p.884).

The situation was even more dramatic since *direct experimental evidence* in favor of the existence of light quanta was almost absent. It famously began to transpire only circa 1923 (the Compton effect).

Being taken independently, the STR did not explain any *new* experimental fact. For obvious reasons (the second STR postulate), the predictions of the Lorentz theory were identical to that of the STR, so that it would not be possible in any case to distinguish the two rival theories on experimental grounds. It is no wonder that most of Einstein's contemporaries had vividly discussed the 'Lorentz-Einstein electron model', brooded on the 'principle of relativity of Lorentz and Einstein', and so forth. At the time of publication of Lorentz's dainty second-order theory (1904), the only data available to test these theories were Kaufmann's dubious measurements of the masses of slowly moving electrons. However, they were initially interpreted by the scientific community as contradicting *both* STR and Lorentz's theory. Thereafter, it took a whole year for Einstein to retort Kaufmann's paper. It is easy to imagine how the STR was accepted by the scientific community in 1905 – 1906.

Furthermore, Einstein prudently did not promulgate the inseverable connections between 1905a and 1905d until 1909. However, without these links, the STR bizarre postulates can be branded *as ad hoc hypotheses*. Surely they were! (The rebuke of Henri Poincaré and the French school is the most apparent example). So, being confronted with many verisimilar rival theories, why did Einstein prefer namely the STR? What indisputable advantages did it possess over the fine artful contrivances of Lorentz, Ritz, and others?

The judicious answer again leads one to Einstein's intricate and distinctive unificationist methodology. It is namely the unificationist stand that excuses Einstein's seemingly puzzling remarks that despite the underdetermination of theory by 'facts' at any given time there is only one correct theory: the theory with the *greatest power of unification* at that time (Einstein 1918; see Beller 1981).

We are usually told that in inventing the STR Einstein had doggedly contrived a *'theory of principle'*, rather than a *'constructive theory'*. However, actual events are not that simple. Forsooth, Einstein himself eminently molded a celebrated distinction between *'theories of principle'* and *'constructive'* ones. Constructive theories strive to 'build up a picture of the more complex phenomena out of the materials of a relatively simple formal scheme from which they start' (Einstein 1919 as quoted in van Dongen 2010, p.49). A manifest example of a constructive theory is the kinetic theory that tries to reduce mechanical and thermal properties of gases to chaotic movements of molecules; the second apparent example is Einstein's bold light quanta hypothesis.

On the other hand, principle theories start not from hypothetical constructions, but rather from empirically ascertained general principles.

"Thus the science of thermodynamics seeks by analytical means to deduce necessary conditions, which separate events have to satisfy, from the universally experienced fact that perpetual motion is impossible". In explicitly Kantian terms, Einstein in 1919 makes out between the abovementioned kinds of theories: "principal theories employ the *analytic*, not the *synthetic* method" (quoted from van Dongen 2010, p.50).

It is to his boon companion Michele Besso that Einstein dedicated the only acknowledgment in his STR paper, the paper that stands out for its lack of any reference to the literature. Nay, in the 1905d paper "the failure of attempts to detect a motion of the earth relative to the 'light medium'" is taken as evidential support only for *one* of the two STR postulates – for the celebrated 'Principle of Relativity'. The odd 'Light Postulate' is alluded almost parenthetically, without any downright discussion of its experimental grounds. Only in subsequent 1905e paper, while recapitulating the 1905d results, Einstein drops an assuaging phrase: "the principle of the constancy of the velocity of light used there *is of course* contained in Maxwell's equations" (Einstein 1989, p.172). Yet one should take into account that for him the 1905d paper was only a provisional product, a mere milestone in implementing the ambitious and farsighted reconcilement program. Einstein was aware that

"A physical theory can only be satisfactory if its structures are composed of elementary foundations. The theory of relativity is just as little ultimately satisfactory as, for example, classical thermodynamics was before Boltzmann had interpreted the entropy as probability "(Einstein to Arnold Sommerfeld on 14 January 1909; quoted from Stachel 2000, p. 10]).

Therefore, the hasty proposition that the 1905d paper constituted a theory of principle is merely *half* of the truth. The 1905d theory was only a constructive one that diligently *posited itself*, showed itself off as a theory of principle. (Probably due to the tactical reasons for Einstein strived to save the STR from the scathing criticism directed against the light quanta). That is why two years later, trying to manifest the STR foundations to the broad physical community, Einstein mildly propounded his relativity theory as merely

"An attempt to summarize the studies that have resulted to date from the *merger* of the H.A.Lorentz's theory and the principle of relativity" (Einstein 1907, p.253).

However, the uncertain position could not last over a long period. Einstein was waiting for the right moment to throw his cards up and to lay out the fidgeting link between his 1905a and 1905d brave innovations. Also, already four years later, in 1909, in Salzburg, he made an overview solid report at the 81-st meeting of German Natural Scientists and Physicians under the promising title '*On the Development of our Views on the Nature and Structure of Radiation*'. Incidentally, it was practically the first effort and opportunity to furnish almost all his multifarious papers as a whole, as well as one of the first Einstein's public reports dedicated to recounting the STR foundations.

The report commences with a succinct recapitulation of luminiferous ether theory that ends with an intriguing question: 'However, today we must regard the ether hypothesis as an obsolete standpoint'.

Why? What was evidently behind all this? What I want to stress is that *for the definite answer Einstein appeals not to the Michelson-Morley* or Fizeau experiments, but contends that

"It is even undeniable that there is an extensive group of facts concerning radiation that shows that light possesses certain fundamental properties that can be understood far more readily from the standpoint of Newton's *emission theory of light* than from the standpoint of the wave theory. It is, therefore, my opinion that the next stage in the development of theoretical physics will bring us a theory of light that can be understood as a kind of *fusion* of the wave and emission theories of light"(Einstein 1909, p. 379; my bold italics).

Besides, the abovementioned textbook experiments are brought into consideration only in the humble context of the "cardinal aspect in which the electromagnetic theory agrees with, or, more accurately, *seems to agree* with the kinetic theory" (Einstein 1909, p. 379; my italics).

3.6 Part five. Concise Discussion.

The thrust of the chapter consists in that to comprehend the important facets of Einstein's 1905 epoch-making creativity and all his 'annus mirabilis' distinctive writings *as a whole*, as well as to fathom the fancy *order* of their presentation one should resort to Planck and Einstein's ardent efforts to reveal the fissure between the profound and entrenched research traditions of classical physics and to commence to reconcile them. In other words, the crux of the present chapter is to buttress the epistemic mellow theory-change model punctiliously displayed in Chapter I.

Briefly, the epistemological import of the lucid model consists in the following. *Profound breakthroughs* in science were chiefly due not to ingenious inventions of new paradigms or bold creation of new daring ideas *ex nihilo*, but rather consisted in the long-term groping and harrowing processes of the *reconciliation, interpenetration, and inter-*

twinement of 'old' research traditions preceding such breaks. Surely, the epistemic model, in its turn, is grounded on the corresponding entrenched epistemic traditions. Just to quote, to begin with, Pierre Duhem,

"In the course of this long and laborious birth, we can follow the slow and gradual transformations through which the theoretical system evolved; but at no time can we see a sudden and arbitrary creation of new hypotheses. A physical theory is not the sudden product of creation; it is the slow and progressive result of evolution" (Duhem 1906/1954, p.252, p.221).

Furthermore, Duhem highly praised Helmholtz's electromagnetic theory for its advances provided 'without losing any of the recent victories of electrical science; it reestablishes the *continuity of tradition*' (Duhem 1902, p.225).

Incidentally, such a reputable researcher as an outstanding editor of "*Annalen der Physik*", Sehr geehrter Professor Max Planck, had proffered in the prophetic paper "*The Relation between Physical Theories*" the first draft of the abovementioned epistemic model (see the epigraph to the present chapter). Moreover, as the prime example of the fine epistemic scheme, Planck scrutinized the problem situation in physics of the second half of the XIX-th century with mechanics, electrodynamics, and thermodynamics as its pivotal research programs. On constituting the cross-contradiction between the mature research traditions of classical physics, Planck proceeds in unfolding the efficacious means of its resolution stating that the impulse of the transformation of mechanics came *from without*, viz. from classical electrodynamics.

"To-day, it is generally accepted that an absolutely rigid mechanical theory of ether consistent with Maxwell's simple equations cannot be countenanced and this makes *a gap between classical mechanics and electro-dynamics* that cannot be bridged. It only remains to define exactly the range of applicability of the two theories or to modify one of them. The former soon proved to be impracticable, for mechanics and electro-dynamics *overlapped* on the question of motion of electrons and the manner of settling the controversy was shown for the first time by the discovery of deviations from the laws of classical mechanics, the deviations finding expression in the variable inert mass of the electron.

Einstein's [special] theory of relativity contains a simple, complete solution of the problem of reconciling mechanics and electro-dynamics in a quite general way, in so far as *it retains the practical essentials of the classical theory* and still fulfills the demands of electro-dynamic calculations. The modification in mechanics due to the principle of relativity is the introduction of a new universal constant, entirely foreign to classical mechanics, namely the velocity of light in an absolute vacuum" (Planck [1908],1960, p. 85; my italics).

Note that in the same magnificent paper Planck indicates the link connecting the STR and the bold light quanta hypotheses through the vicissitudes taken by dynamics in its accommodations to the Maxwellian electrodynamics.

"However, it appears that this is not *the only*, nor the most difficult, a sacrifice which dynamics must offer if it wishes completely to embrace the theory of heat. It has probably not finished with the question of discontinuity of matter. The laws of heat radiation, specific heat, electron emission, of radio-activity, and yet many other branches are in agreement that not only matter itself but also the effects radiated from matter ... possess discontinuous properties, which again can be characterized by a new natural constant: the elementary quantum of action" (Planck [1908],1960, p.85).

Similarly, in contriving the indispensable theories of light quanta and STR Einstein was operating according to a solid belief in the necessity for unity in science, as well as the coincidence of both constructing theory and integrating intuition of Principle. Whereupon, sagacious identifying the paradox caused by the profound contradictions between the pivotal research traditions and dexterous resolving it ('must be modified to remain compatible') turns out an *ineradicable part of Einstein's scientific method.* Besides, to comprehend the importance of the latter one should invoke Mach's renowned principle of the economy of thought that governed Einstein's consciousness through and through his entire dizzying scientific career. Nevertheless, this is not to allege that 1905 Einstein was an obdurate Machian incapable to draw upon the rival epistemic sources.

Nay, for instance, in profound *"Physik und Realität"*, published in 1936, i.e. in a conspicuous paper that represents the acme of his philosophical activity, Einstein takes the history of mechanics in quite the opposite from the author of "*Die Mechanik*" way (Einstein [1936], 1954, p.302). It is alleged that the fundamental divergences of opinion with Mach sprung out from the stubborn development of atomic theory by Einstein through 1905 inquiring into Brownian motion (Einstein 1905b). Hence, the advancement of a similar idea of 'atoms of light' (Einstein 1905a) can be taken as a part of the abovementioned research activity. Just compare this litigation with Einstein's attitude to Ludwig Boltzmann exposed in the September 13, 1900 letter to Mileva Marić:

"The Boltzmann is absolutely magnificent. I'm almost finished with it. He's a masterful writer. I'm firmly convinced of the correctness of the principles of the theory..." (CPAE, vol.1, Doc.75; translated by Renn and Schulman 1992, p.32).

Yet, in my judgment, to comprehend the more profound reasons of the abovementioned divergences, one has to turn face to face to Einstein's highly debatable and controversial overall philosophical creed.

Overall, Einstein's distinctive standpoint can be characterized as 'sophisticated *eclecticism*'. Besides, one cannot elude peering at the famous passage from Einstein's epoch-making 1949 "*Reply to Criticism*" where he openly acknowledges his overall eclectic standpoint (Einstein 1949b, p. 684). Hence it is no wonder that I strongly support Don Howard's felicitous claim that more thoroughly Einstein's philosophy of science can be professed as a *quaint fusion* of the elements drawn from sources as diverse as 'Machian empiricism, Duhemian conventionalism, and neo-Kantianism' (Howard 1994). Yet, what is the mutual arrangement of the components of this whimsical mixture?

Firstly, the indubitable influence of Machian empiricism on 1905 Einstein's creativity was envisaged earlier. Yet, in a famous letter to Michele Besso of January 1948, Einstein confides that

"I see his [Mach's] great service as residing in the fact that he dispelled the dogmatism that reigned in the foundations of physics in the 18th and 19th centuries. Especially in the "Mechanik" and the "Warmelehre", he sought to show how concepts grow out of the experience. He convincingly defended the view that these concepts, even the most fundamental ones, obtain their justification only from experience... I saw his weakness as residing in the fact that he more or less believed that science consists in the mere "ordering" of empirical materials; i.e. he misunderstood *the free, constructive element* in the formation of concepts. In a sense, he believed that scientific theories arise through discovery and not through invention" (Speziali 1972, pp. 390-391; my italics).

Secondly, as for (often obscure and obliterated) neo- Kantianism, its influence on 1905 Einstein is not limited by the regulative principles of epistemology. The very 1905a light quanta hypothesis is a *constructive* model of radiation; hence, in the 1949 *Autobiographical Notes* Einstein recalled of Mach's controversial legacy:

"He [Mach] did not place in the correct light the *essentially constructive* and speculative nature of all thinking and more especially of scientific thinking; in consequence, he condemned theory precisely at those points where its *constructive-speculative* character comes to light unmistakably, such as in the kinetic theory of atoms" (Einstein 1949a, p. 13; my italics).

The hard fact that the constructive nature of light quanta hypothesis brings Einstein's thought closer to Kantian epistemology was already punctuated by many Einstein scholars (see for instance Lenzen's fascinating comments in Schilpps' eminent 1949, p.380). Furthermore, in an eloquent letter to Moritz Schlick from November 1930, Einstein famously alleged:

"Physics is an attempt to construct conceptually a model of the real world as well as its law-governed structure" (quoted from Howard, 2014, p.25).

Respectively, the aspiring drive for creative work could be elicited in Kant's *constructivist* foundation for scientific knowledge that restricted science to the realm of appearances professing that *a priori* knowledge of things in themselves is impossible. Much later Einstein pointedly confessed:

"I did not grow up in the Kantian tradition but came to understand the truly valuable which is to be found in his doctrine, alongside of errors which today are quite obvious, quite late. It is contained in the sentence: 'the real is not given [gegeben] to us, but put to us [aufgegeben]' (Einstein 1949; quoted from Ryckman 2005).

Even mathematics – asserted to be most stout and certain because of its being analytical – was conceived by Kant as an *a priori synthetic judgment*. As he famously enunciated in *"Prolegomena"* (Kant [1783], 2002), the asset of pure mathematical cognition, differentiating it from all other a priori cognition, is that it must advance not from concepts, but through the *construction* of concepts. Pure mathematical cognition, in its abstract propositions, must, therefore, go beyond the concept to that which is contained in the *intuition* corresponding to it. Thus, its propositions can and must never spring out of the analysis of concepts, i.e. analytically, and so are all-synthetic.

The Kantian thought-provoking doctrine of the intuitive nature of mathematics means the limiting of mathematics to those objects that are constitutable [Konstruierbar]. 'Intuitive' is tantamount to 'constitutable'. As Ludwig Wittgenstein has later elicited in genuine Kantian terms, 'But the mathematician is not a discoverer, he is an *inventor*'.

Kant regarded objectivity of science as resulting from the relentless way in which the turmoil of sensibility was punctiliously arranged under the solid categories of the understanding with the help of spatial and temporal categories. That is why mathematics can so effectively highlight objective reality: for Kant mathematical constructs are related to the pure intuitions of space and time.

Hence, mathematical statements are true in virtue of their application in experience to posit the behavior of empirical bodies. While mathematical judgments are rendered via construction in pure intuition, they are taken as cognitions only because they are inextricably connected to experience in the sense that geometrical space was contemplated as a necessary condition of appearance (Kant [1787], 1998, p.196).

Since numerous abstract objects of a theory are composed by the laws of the theory, the objectivity of mathematics is constituted not by the existence of things themselves but by the *objective validity of relations*. Respectively, in the 1905a masterpiece, while carefully constructing the mathematical abstract object 'light quanta' out of the basic objects of Maxwellian electrodynamics, statistical mechanics, and classical thermodynamics, Einstein was not bothered with grasping the inextricable 'essences' of radiation phenomena. He grappled with the problems of *reconciling* the interrelations of the abovementioned classical physics research traditions. Incidentally, in their beneficent proposal for Einstein's membership in elitist Prussian Academy of Science, M. Planck, W. Nernst et al. had stressed that

" Einstein has a special talent for getting to the bottom of other scientists' newly emerging views and assertions, and for assessing *their relationship to each other* and to experience with surprising certainty" (Doc. 445 of Einstein 1987, p.338; my italics).

As is well-known, Einstein's philosophical evolution after the GTR was achieved unalterably carried him further and further from Humean and Machian half-digested empiricist bias. It led him toward vigorous and profound neo-Kantian tradition represented by such luminaries as Hermann Weyl, Arthur Eddington, Ernst Cassirer, Edmund Husserl, et al., and the corresponding mathematical speculative methodology embodied in a sequence of unified theories. Thus, I do not contend that Einstein of 1905 was an obdurate (neo) Kantian, vehemently trying to implement the abstract and murky dictums of "Critique" (almost unfathomable to common folks) into his everyday research practice. Yet, in my view, *the Kantian roots* (which need to be trialed by future research) of Einstein's subsequent efforts find their embodiment in the 1905 research activity indissolubly connected with his fruitful attempts to reconcile Maxwellian electrodynamics and statistical thermodynamics.

Furthermore, despite the dominant and profound (yet not fully realized by young Albert - owing to youthful romantic ideas and revolutionary mood – Feuer 2017) influence of Kantian epistemology, one should not ignore Einstein's simultaneous propensity to Pierre Duhem's distinctive brand of conventionalism. Though Duhem's epoch-making and favorably received "*The Aim and Structure of Physical Theory*" first appeared only in 1906 (and the second edition in 1914), two Duhem's articles, expounding many pivotal ideas of the book, appeared in the "*Revue de Philosophie*" already in 1904-1905. The crux of Duhem's startling book eminently consisted in the assertion that

"A physical theory is *not* an explanation. It is a system of mathematical propositions, deduced from a small number of principles, which aim to represent as *simply*, as *completely*, and as *exactly* as possible a set of experimental laws"(Duhem [1906], 1954,p.19; my italics).

As for Albert Einstein, when, in the fall of 1909, he left his salutary job at the meager patent office in Bern and returned to Zurich to take up his conspicuous academic position at the University of Zurich, he happened to rent an apartment at Mousonstrasse 12, directly upstairs from his old friend Friedrich Adler. It is quite important that just one year earlier Friedrich published his German translation of Duhem's "*La Theorie Physique: son Objet et sa Structure*". In one of the distinctive letters of this time (1909, to Victor Adler) Friedrich confided that

"The more I speak with Einstein – and that happens fairly often – the more I see that my favorable opinion of him was justified. Among contemporary physicists, he is not only one of the clearest, but also one of the most independent minds, and we are of one mind about questions whose place is generally not understood by the majority of other physicists" (quoted from Don Howard, 2014, p.7).

Einstein's substantial immersion into Duhem's holism is faintly indicated by the 1910-1911 "*Lecture Notes for Course on Electricity and Magnetism*", the University of Zurich, winter semester 1910-1911, where he operates freely with inherent Duhemian notions of the 'totality of experience', and so on (Don Howard, 2014). Yet, in my view, one can proceed further stating that for Einstein Duhem's conventionalism was a *solid bridge* indispensable to connect and accommodate two disparate subjects of Machian crude empiricism and Kantian academic apriorism and transcendentalism. For instance, in the Appendix to "*The Aim and Structure of Physical Theory*," Duhem states as 'the great fact summarizing the whole history of science' the diversity fusing into more comprehensive and more perfect unity. Thus if a physicist

"yields to the nature of the human mind, which is repugnant to the extreme demands of positivism [...] he will note that physical theory through its successive advances tends to arrange experimental laws in order more and more analogous to *the transcendental order* according to which the realities are classified, that as a result, physical theory advances gradually toward its limiting form, namely that of *natural classification*, and finally that *logical unity* is a characteristic without which physical theory cannot claim this rank of a natural classification"

(Duhem [1906], 1954, pp.296-297; my italics).

Furthermore, the Appendix concludes by a rhetorical question

"Does theory tells us something concerning reality which experiment has not taught us and could not possibly teach us, that is, something *transcending* empirical knowledge? If we must answer this last question affirmatively, we shall be able to say that physical theory is true and that it has value as knowledge" (Duhem [1906], 1954,p.326).

Yet in an already-quoted letter to Max Born Einstein unwinds the subtle relations between Kantian and conventionalist epistemologies in the following sincere way:

"I am reading Kant's "*Prolegomena*" here, among other things, and am beginning to comprehend the enormous suggestive power that emanated from the fellow and still does. Once you concede to him merely the existence of synthetic apriori judgments, you are trapped. *I have to water down the "a priori" to conventional,* so as not to have to contradict him, but even then the details do not fit" (Born 1969, pp. 25-26; my italics).

Hence, Duhemian discreet conventionalism helps to reconcile Kant's sublime apriorism and Mach's mundane empiricism.

To recapitulate, Einstein was influenced by Hume, Mach, Kant, Mach, Poincaré, Duhem, Hertz, Helmholtz, et al., and this is evinced in multifarious documents – papers, letters, lectures, oral communications, etc. - relating to diverse periods of his magnificent creativity. Nevertheless, if one delves mainly into his scientific papers and notebooks, craving to grasp, according to Einstein's shrewd recipe, his *modus operandi,* one finds out sober reasons to believe that, at least in 1905, in his mundane research practice, he had held such an epistemological stand that can be characterized as 'sophisticated eclecticism'. The latter constituted a quaint but discreet fusion of Machian, Duhemian, and Kantian intricate epistemologies. And the most substantial Machian concept indispensable to comprehend Einstein's 1905 research activity as a whole is Mach's principle of economy of thought yet blandly taken in the context of intuitive knowledge doctrine. Hence, Einstein renounced the notorious ether notion not because it was a metaphysical straw man, an idle concept, an antediluvian superfluous contraption. He eliminated it first and foremost since it turned out a flagrant blemish for efficacious reconcilement of Maxwellian electrodynamics and statistical thermodynamics that promised to pave a gallant highway to the stupendous quantum field theory and – ultimately – to general relativity.

4 ADVANCEMENT OF GENERAL THEORY OF RELATIVITY: THE DAINTY RECONCILIATION OF THE PARTIAL THEORETICAL SCHEMES OF ABRAHAM, EINSTEIN, AND NORDSTRÖM

"Both the peculiar emergence and the remarkable stability of Einstein's theory of gravitation with regard to the further development of physics and astronomy becomes plausible only if the genesis of general relativity is understood not as a fortunate anticipation of future observational discoveries, but as a transformation of pre-existing knowledge."

Michelle Janssen, John Norton, Jurgen Renn

Abstract.

The aim of the chapter is to amend the received view on the general theory of relativity (GTR) advancement owing to common scientific practice of its application and functioning, the history of science 'stubborn facts' and the philosophy of science sober reasons. The origin of GTR as an instance of lucid epistemic model of mellow theory change that hinges upon 'old' mellow theories encounter and intense interaction is doggedly elucidated. The arguments in favour of the strong tenet that the dynamic creation of the GTR had been continually governed by profound internal tensions between two profound research traditions, that of special relativity and Newton's gravity are strengthened. The 1907-1911 encounter of the traditions, their deep

150

interpenetration and subtle intertwinement entailed construction of vast hybrid realm at first with haphazard throng of crossbred theoretical models. Step by step, on consecutive soothing and eliminating the contradictions between the models contrived, the hybrid set was arranged by dint of the celebrated 'Principle of Equivalence'. It is contended that one of the reasons for the GTR victory over the deft rival research programmes of Abraham and Nordström was a substantially synthetic character of Einstein's program. As a result of efficacious reconciling and amalgamating the solid 'physical' and 'mathematical' approaches, embodied in Abraham, Einstein and Nordström's crossbred partial theoretical models, Einstein was able to explain away at last the anomalous motion of Mercury and to triumph over the rivals.

Keywords: Einstein, Abraham, Nordström, reconciliation, crossbred models, general relativity, Entwurf, nonmetric theories.

4.1 Introduction. Sad Fate of Abraham and Nordström's Fine Theories.

It is well-known that Albert Einstein's vehement efforts to yield the General Theory Relativity (GTR) were accompanied by its rival versions deftly contrived by Gunnar Nordström, Max Abraham, Gustav Mie et al. In particular, in 1912-1914 a skilful Finnish mathematician Gunnar Nordström advanced a scalar Lorentz covariant gravitation theory.In a conspicuous paper submitted to renowned '*Physikaliche Zeitschrift*' in October 1911 Nordström alleged that he had elicited a wilful rival to Einstein's conspicuous hypothesis which would "...leave c [the speed of light] constant and still adapt the theory of gravitation to the relativity principle in such a way that gravitational and inertial masses are equal" (Nordström 1912, p. 1126).

Likewise, in 1911 a Gottingen celebrated master of classical electrodynamics professor Max Abraham furnished a host of scalar and vector Lorentz covariant dainty gravitation theories where light and gravitation transpired to possess the same speed of propagation. In Abraham's skilful scalar theories "c, the speed of light, depends on the gravitational potential. This hypothesis was first enunciated by Einstein (Ann. d. Physik, 35, 1911, p.898)" (Abraham 1912b, 793).

The fine papers of Abraham and Nordström are still considered as whimsical *delusions* capable to stir up problem situations at best and to incite critical discussions around the GTR queer genesis highlighting all its magnificent opulence and dazzling splendor. Eventually a solid received view on the origin of the GTR was set up according to which, in the process of its sedulous creation, Einstein decisively 'rejected' Lorentz covariant scalar and vector theories of gravity.

Nevertheless, certain current history-of-science insights (Norton 1992; Renn and Sauer 2007; Renn 2007b; van Dongen 2010) impel one to take the standard view with a substantial grain of salt. To begin with, the Einstein - Nordström downright correspondence convincingly reveals that it was Albert Einstein himself who, before November 1915, and even *after* the creation of GTR preliminary metric version – the notorious '*Entwurf*" (1913) – took active part in working out Nordström's artful scalar relativistic theories. Einstein was in close contact

152

with Finnish assistant professor during the very period in which the Nordström theory was developed. The theory actually advanced via a vigorous exchange between Einstein and Nordström, with Einstein often generating distinctive ideas decisive to development of the theory. By and large the theory might more accurately be called the 'Einstein-Nordström theory'.

The next suitable example is Albert Einstein and Adrian Fokker's thought-provoking paper published in early 1914 that aimed at no more and no less than 'application of new mathematical methods, used in Einstein and Grossmann's paper, to Nordström's theory' (Einstein and Fokker 1914). Incidentally, in the same paper in early 1914 the striking connections between Nordström's artful theory and conformally flat spacetimes were discovered. Therefore it was within Nordström's fine research program where the gravitational field equation

$R = \chi\, T$ (χ = const) was first derived (1912), with R being fully contracted Riemann-Christoffel tensor and T the trace of the stress-energy tensor (in an important case of an unstressed, static matter distribution). The dainty field equation is an apparent harbinger of Einstein's celebrated equations presented to Preussiche Akademie der Wissenschaften on November 25, 1915 (Einstein 1915).

Furthermore, the important consequences of the '*Entwurf*' and the GTR coincide with the corollaries of the theories of Nordström and Abraham for a number of vital cases in certain reasonable approximations. For instance, the '*Entwurf*' is reduced to a theory with a four-vector field potential that is formally analogous to maxwellian electrodynamics in the so-called 'weak - field approximation'. Moreover, special theory of relativity (STR) turns out to be an indispensable intermediary step in the thorny transition from the GTR to Newton's pre-eminent theory of gravitation (see Landau and Lifshitz 1987 for details). Though this disputable transition is grounded on the indispensable premise, for weak and stationary gravitational fields, that the gravitational field is described by a scalar in *flat* (Minkowski) spacetime, i.e. on the inevitable reduction to scalar Nordström's theory.

Likewise, the so-called 'linear approximation' in GTR, still in common use to account for gravitational waves (Einstein 1916) origin, propagation and detection (see, for instance, the discussion around recent 2015-17 LIGO pre-eminent experiments), presupposes the transition to such a theory of gravitation in which a gravitational wave, in full analogy with classical electrodynamics, is described by a vector in flat space-time, i.e. the transition as a matter of fact to vector theory of Abraham (1915). Abraham's ambitious claim that his theory contained Einstein's as a limiting case was rebutted by Einstein, though. But, nevertheless, in the both cases the intricate relations between mellow GTR and deft theories of Nordström and Abraham strongly resemble the common pattern of classical electrodynamics where the general potential is represented by multi-component object such as a vector or a tensor, which, in the special juncture of a static field, reduces to a single-component mathematical object.

Furthermore, Einstein's initial approach to the GTR contrivance, - on the one hand, - and Nordström's and Abraham's enticing approaches, - on the other hand, - where in many crucial aspects **complementary**. For instance, Einstein's wayward theory of static gravitational field (Einstein 1912a) was incited by substantially physical considerations based on the Equivalence Principle, while Abraham's dainty theories prudently started from mathematical considerations related to Minkowski formalism. Yet the punctilious investigation of the so-called '*Zurich Notebook*' that was amply employed by Einstein while constructing the '*Entwurf*' theory (where the corresponding entries begin in mid-1911 and end in early 1913) brings some astonishing novel light on Einstein's thought laboratory.

Investigation of the '*Zurich Notebook*' has elicited that Albert Einstein was persistently struggling forward along the so-called '*dual strategy*' that embraced both a physical and mathematical approaches to finding the ultimate field equations (Janssen et al, 2007; Renn and Sauer 2007; van Dongen 2010). The discreet 'dual method' suggested that Einstein proceeded first and foremost from a set of constraints of a *physical* nature (Newtonian apparent limit plus ineradicable conservation of energy and momentum). On the other hand, the complementary 'top-down' (van Dongen) and as plain as a pikestaff *mathematical* approach originated from no less judicious Principle of General Covariance.

The key point consists in that the 'dual strategy' necessarily employed *both* physical and mathematical approaches: "it was an *iterative* process that began with trying out one approach, and then checking results with the demands or results of the other approach" (van Dongen 2011, p. 11). Physics-first prevailed but lead to notorious '*Entwurf*'; though then the mathematics approach took over and incessantly led to the full-blooded 1915 General Theory of Relativity.

Nevertheless, it should be stressed that Einstein's judicious 'physical' and 'mathematical' strategies, the vacillation between whom brought to '*Entwurf*' and GTR poignant contrivance, were entrenched in two rival research traditions encountered circa 1907 (Renn and Sauer 2007, p.125).

Einstein repeatedly made clear his preference for Nordström's spellbinding theory over other artful rivals; for instance, in September 1913 presentation of the '*Entwurf*' theory (see section 3 for details) to the 85th Congress of the German Natural Scientists and Physicians in Vienna. His single and rather meek critical remark consisted in that the theory was regrettably incompatible with Mach's principle – a vice that could turn out a virtue to a Naturforscher feeling real animus against whimsical metaphysical castles in the air. Later none other than Wolfgang Pauli (1921) hastily christened Nordström's theory an 'empirical blunder' since it had not predicted any deflection of a light ray by a gravitational field and had not properly explained the anomalous motion of Mercury. Yet there had been no eclipse expeditions in 1913 and Einstein and Grossmann's '*Entwurf*' turned out to be a hangdog incapable to provide a trustworthy explanation of the exasperating anomalous motion of Mercury.

On the contrary, in 1911 G. Pavani meticulously calculated the perihelion shift of Mercury according to Abraham's renowned theory, arriving at a sordid value of 14", 52, that is approximately only one third of the observed one. Alas, even Abraham's vector theory made a more accurate prediction than the notorious 'Entwurf'. If one allows that Einstein's discovery (due to Michele Besso's diligent calculations) of the failure of the '*Entwurf*' theory to yield the correct perihelion shift of Mercury was furnished as early as the summer 1913, one can conclude that Abraham's advancement was a vigorous spur that prompted Einstein to construct GTR out of the '*Entwurf*'.

All the above indicated hallmarks of the GTR origin, advancement and common functioning, and preponderantly *the common practice of its implementation* impel to draw the following peculiar conclusions.

1. The intricate relations between the GTR and its deft rivals were far more complicated in 1907-1915 than it may seem from the pestered 'truth-falsity' dilemma, so that one can with surprise contemplate the interlacement and *interpenetration* of rival 'incommensurable paradigms' into each other.

2. Einstein's mellow GTR was better than its inimical rivals if only for the reason that it *encompassed them all* though in significantly modified forms. (Just as the GTR blandly embraces Newton's theory of gravitation, as well as the STR, or just as the maxwellian electrodynamics famously encompasses the partial theoretical schemes of Coulomb, Ampére, Biot & Savare et al.).

3. Einstein could accomplish the reconciliation of the knowledge on gravitation and inertia (rendered by classical mechanics) and the knowledge on the structure of space and time (embodied by the STR) via the painful 'Entwurf – GTR' transition only. As a result of reconciling and amalgamating the 'physical' and 'mathematical' approaches, embodied in Abraham, Einstein and Nordström's crossbred partial theoretical models, Einstein was able to explain away the exasperating anomalous motion of Mercury.

Hence *the ultimate aim of the present chapter* is to amend the standard view on the GTR origin and advancement by taking into consideration the excavated history of science 'hard facts', philosophical (and sociological) solid reasons and modern common scientific practice of its functioning. My *main idea* consists in that one of the reasons for the GTR dazzling victory over the rival programmes of Abraham and Nordström lied in a substantial synthetic character of the Einstein research program. Einstein's farsighted program did supersede the rival ones because it did deftly assimilate sober premises of the Nordström fine program as well as judicious presuppositions of the mature program of Abraham. In particular, the Einstein program's convincing victory over its rivals became possible since Einstein had put forward as a pivotal synthetic principle the Principle of Equivalence that radically differed from that of rival approaches by its open, flexible and contra-metaphysical, regulative character.

In the *second section* of the paper the relevant facets of chapter 1 lucid epistemological model that fits some achievements of current philosophy & history of science and dwells on mellow theory dynamics and structure are recollected. The epistemic model is a gist of the present study contending that the history of physics does not advance through the creation *ex nihilo* of novel paradigms, but rather through tantalizing, long-term and groping processes of reconciliation, interpenetration and intertwinement of 'old' entrenched research traditions.

In the *third section* of the chapter the initial stage of the GTR contrivance (1907-1912) is scrutinized. The crux is a premise that the invention of relativistic theory of gravity had actually commenced through the crossbred object construction in Einstein's 1907 farsighted paper, i.e. with the implantation of mass-energy relation into the theory of gravity. The crossbred object entry – the introduction of inertial and simultaneously gravitational mass – led to a penetration of STR subtle methods into Newtonian theory of gravity and to a reverse penetration of Newtonian gravity methods into the STR. As a result, the both theories were radically rebuilt from within and the corresponding changes in both of them were manifested. The changes were epitomized in the specific sequences of crossbred models, the byproducts of the transformation performed.

1. On the one hand, an inevitable corollary of relentless penetration of STR into Newtonian theory of gravity turned out to be Nordström's and Abraham's scientific research programs.

2. On the other hand, no less inevitable, owing to felicitous Equivalence Principle, was the Newton theory intrusion into the STR. It led to the sequence of Einstein's advances on the generalization of relativity principle and to spreading the principle not

only on inertial systems of reference, but on the multifarious accelerated systems as well. However, the most promising result of the hybrid theories of Nordström and Abraham consisted in that both deft constructs manifested some *extremely promising assets* on how the global theory could be put up. As a result, the GTR sprung out as a substantially 'synthetic account' artfully combining all the positive achievements of the other rival approaches. Hence the climax of the stage was Einstein's bold proposal and apprehension of magnificent equivalence principle that became one of the firm GTR heuristic foundations.

The *fourth section* of the chapter (1912-1913) is dedicated to the '*Entwurf*' jogging construction. This metric theory was put up not as 'phoenix from the ashes' but via the meticulous synthesis of Abraham's and Nordström's theoretical schemes, as well as from the preliminary nonmetric theoretical schemes of Einstein himself. The staple was the metric tensor introduced owing to the Equivalence Principle and via Nordström, Laue and Planck's startling results.

And, finally, the *fifth section* of the paper (1913-1915) grapples with the GTR construction out of the "*Entwurf*". It is stressed that the main thrust in passing from the '*Entwurf*' to the GTR came from Einstein's farsighted 'dual strategy' rousing him to *reconcile* physical and mathematical approaches. And only *after* he blandly reconciled them. i.e. only after he comprehended the general covariance of the gravitational field equations could he turn to successful explanation of the exasperating Mercury perihelion. Hence the decisive impetus in constructing the General Theory of Relativity came from the craving attempts to reconcile physical and mathematical approaches, embodied in Abraham, Nordström and Einstein's crossbred partial theoretical models. Trustworthy explanation of Mercury perihelion motion transpired as a lucky by-product of the strenuous reconciliation efforts.

4.2 Part one. The Subtle Structure of Modern Mellow Theory of Gravity.

The flustering dynamics of the theory of gravity was predominantly governed by strong *internal tensions*, blatant contradictions *within* the knowledge system rather than by new empirical knowledge, which played only a subordinate role at best. In this section, some relevant to the theory of gravity features of a corresponding perspicuous epistemological model dealing with mature theory dynamics and structure are recollected.

The current philosophy of science argle-bargle discourse on scientific revolutions allows one to elucidate the views on the structure and functioning of scientific theories, – on the one hand, - and to contrive sufficiently sweeping and exact theory change epistemic models, – on the other. In particular, according to one of the models elicited in chapter 1 (see also Nugayev 1999, 2018), a scientific revolution is provoked by ineradicable encounters of some entrenched 'old' paradigms, scientific research programs or research traditions that cannot be matched in a common way – by reducing one of them to another. The most efficacious and radical way out of the predicament is to work out such a global theory that encompasses all the theories involved in significantly modified forms. The global theory is aimed at eliminating the tensions and smoothing away frictions between diverse paradigms implicated.

In the troublesome course of global theory contrivance an indispensable preliminary stage shows itself off in the construction of a series of intricate hybrid theories. The latter are persistently set up to the climax when such felicitous hybrid models are constructed that are able to outline the fruitful way of the global model creation through the wayward generalization of models that belong to the lower level of mature theories. According to the lucid epistemological model, *radical breakthroughs in science sprung out not due to odd invention of new-fangled paradigms or the unfathomable creation of novel ideas ex nihilo, but rather to the long-term groping and relentless processes of the reconciliation and interpenetration of 'old' research traditions preceding such breaks.*

It is a commonplace that no profound epistemic model of scientific revolutions can be properly set up without preliminary elucidating the structure of mellow scientific theories. Yet what I want to punctuate is that a mellow theory of XIX and XX centuries physics encompasses not a single model or a bundle of models. It artfully embraces a *bundle of groups of models* that are related to one another in rather subtle ways. A mature theory is so structured that the throng of its multifarious models is disseminated over at least three following interconnected levels (Stepin 2005; see also the corresponding relevant comments in Vihalemm and Müürsepp, 2007).

1. The *top* level of the **basic** ideal model where 'the Fundamental Theoretical Scheme' resides.

2. The *intermediate* level of the *subordinated* ideal models (or 'the Partial Theoretical Schemes') constructed out of the basic one according to certain (often tacit) rules.

3. The *inferior* level of the *'Empirical Schemes'* that can be approached through the level of partial theoretical schemes, on the one hand, and via the wayward and often disputable 'generalization' of the controversial empirical data, - on the other.

The set of basic objects of a mellow physical theory constitutes the *basis*, i.e. the subsystem of fundamental theoretical objects. All the basic theoretical objects are ultimate idealizations and cannot be taken as substantial bodies. Though the derivative subsystems are *subordinated* (Stepin) to the basic one, they are independent of each other, referring to different fragments of the same domain of validity. Each subsystem is characterized by its own set of notions and mathematical equations that constitute a special part (section) of the mature theory.

In general, the relations between the subsystem of basic objects and a subsystem of derivative ones can be described as follows. A derivative system is obtained from the basis by a process of *reduction*. It means that a mature theory is developed not by formal (logical, mathematical) means only, but also through *gedankenexperiments* with abstract theoretical objects. The reduction is put into effect by taking into account, 'sifting' the features of the empirically fixed domain of validity.

160

This domain can be 'seen through' a cognitive lens of an ideal model, formed by a set of basic objects. According to the peculiarities of each concrete experimental situation, various constraints may be imposed on the system of basic theoretical objects. This enables one to radically transform the system, metamorphosing it into a subsystem of overmastered derivative objects. The fundamental equations are then applied to highlight the properties of the subsystems of derivative objects. In accordance with these properties, they are deftly transformed into the partial laws. The vague and informal nature of such manipulations converts such an inference into an uneasy problem-solving operation. The solutions of such problems are embraced by a theory at its commencement. To a theoretician delved into the application of the theory, they serve as a pattern for subsequent activity. Each problem is solved in accordance with primary 'paradigms' (in Thomas Kuhn's sense).

In the GTR peculiar case, the host of paradigm examples ineradicably embraces the derivation of the laws of Newton's theory of gravity from Einstein's equations in the so-called 'weak field approximation'. When the gravitational field is weak, such a felicitous system of reference can be chosen in which all the metric tensor components slightly differ from their Minkowski values: g = η + h (see, for instance, Landau and Lifshitz 1983). The constraint to ignore the squares and other multiples of h is indispensable for the successful transition to Newton's theory of gravity. But it means nothing else that the index-rising operation is carried out by η – the metric tensor of *flat* space-time. As a result, in the weak- field approximation the gravitational field equations take the form of common wave equations in flat space time for (Nordström's) scalar potential φ. Thus the basic theoretical object of Nordström's nonmetric scalar theory turns out to be constructed from the GTR basis. "This is quite natural since the weak [gravitational] field is taken as a tensor in flat space-time" (Zel'dovich and Novikov, 1973, p.56) and is described by a corresponding felicitous equation

$$h_{ik} = -\eta_{ik}2\varphi/c^2$$

The tidy *construction* of derivative objects from the basic ones enables one to compare theoretical knowledge with experience, to explain the results of real experiments. To this end, an empirical equation - an intermediate relation - is derived from the partial law. In this equation the special constructs are introduced. In contrast to abstract objects, the newly born constructs are no longer idealizations and can be compared with *real* bodies now. These constructs are called *empirical objects*, and their systems - special representations of actual empirical situations - are called 'empirical schemes'. The empirical objects are not tantamount to the real ones. An empirical object cannot be compared with a single body with which an experimentalist operates, but only with a *class* of such objects. Consequently, an empirical scheme corresponds not to a concrete experimental situation, but to a *type of such situations*. For example, the empirical scheme of the Biot & Savare experiment with a magnetic needle and a conductor refers to any experiment with any current in the conductor and any sufficiently small magnetic needle.

A mellow theory becomes an *established* one when the links between all the three levels of the organization are vigorously set up; it makes possible to use the mellow theory as an effective instrument for making definite predictions. The links between all the three levels of an established mellow theory should be sufficiently solid and stout ones. Their rigidity allows one to connect a prediction referring to the upper level with all the underlying levels of a mellow theory. Hence it allows one to construct an experimental device capable to check the prediction. A new result, obtained in the course of advancement of mathematical apparatus, influences all the levels of a mature theory immediately. Hence a mature theory can predict, and the predictions can be successfully verified. A mature theory obtains the status of an established one when at least some of its predictions are posited to be successful. It exhibits that the system of basic objects is complete, and all the necessary links between all the three levels are made sufficiently robust.

Owing to subtle structure of a mellow theory, the global theory contrivance appears to be a slow, adamant and consequent ascent from the lower levels up to the top ones. Any transition from the

lower level to the upper one is impossible until the construction of all
the lower-level models is successfully accomplished. Yet note that the
lower auxiliary models (that served at scaffolding the upper ones) are
not completely eliminated; they can be discovered not only in history-
of-science archives or antediluvian scientific magazines. They can be
transpired in real practice of theories' functioning (in implicit forms,
as a rule). I reckon that the basic models of Nordström's and Abraham's deft theories have turned into the partial theoretical schemes of
the GTR, as well as Pieter Garber's 1898 phenomenological Mercury
perihelion results constitute an empirical scheme of the GTR.

4.3   Part two. Discreet Composition of the Hybrid Models via the Equivalence Principle.

The advent of the special theory of relativity (STR) and the apparent
incompatibility between Newton's theory of gravitation and the STR
perplexed Einstein and his contemporaries with the task of contriving a relativistic theory of gravitation. The profound *contradictions*
between the theories consisted first and foremost in the fact that according to Newton's theory the velocity of gravitational interaction was
ineradicably infinite. On the other hand, the STR famously prohibits
the signals travelling faster than light. Note that undisguised disparity
between the diverse concepts of 'action at a distance' and 'instanta-
neous action' was disclosed just after the construction of maxwellian
electrodynamics. First of all, it was James Clerk Maxwell himself who
tried to invent the first vector theory of gravity. Alas, he was impelled to leave the efforts soon due to the problem of negative energy
of gravitational wave. STR creation only exacerbated the exasperating
problem (see Corry 2004, Petkov 2015 for details).

It therefore comes as no surprise that it was Einstein's 1907 farsighted review "*On the Relativity Principle and the Conclusions Drawn
from it*", published in Johannes Stark's "*Jahrbuch der Radioaktivität
und Elektronik*", that laid the true conceptual foundations for relativistic theory of gravity.

"The most important result of the fourth part is that concerning the inertial mass of the energy. This result suggests the question whether energy also possesses heavy (gravitational) mass. A further question suggesting itself is whether the principle of relativity is limited to nonaccelerated moving systems. In order not to leave this question totally undiscussed, I added to the present paper a fifth part that contains a novel consideration, based on the principle of relativity, on acceleration and gravitation" (Einstein 1907, pp. 254-255).

In the fifth part of his epoch-making 1907 paper Einstein enunciated at the first time his renowned *"Principle of Equivalence"*. As he later recalled, when he had accomplished the 1907 review article for publication, he had tried to modify Newton's gravitational theory so as to reconcile it with the SRT. The corresponding attempts had demonstrated that it was possible, but Einstein had relinquished them since they were grounded on physically inacceptable hypotheses.

"At this point, there occurred to me the happiest thought in my life [der glücklichste Gedanke meines Lebens]. *Just as in the case with the electric field* produced by electromagnetic induction, the gravitational field has similarly only a relative existence. For if one considers an observer in free fall, e.g. from the roof of a house, there exists for him during this fall no gravitational field – at least not in his immediate vicinity. Indeed, if the observer drops some bodies, then these remain relative to him in a state of rest or in uniform motion, independent of their particular chemical or physical nature" (Pais 1982, 178 ; my italics).

Because of the importance of the Equivalence Principle for the creation of GTR and abiding discussions on its 'true content', we have to resort to all the piece of 1907 paper where the principle had been enunciated first.

"We consider two systems $\sum_1$ and $\sum_2$ in motion. Let $\sum_1$ be accelerated in the direction of its X-axis, and let γ be the (temporally constant) magnitude of that acceleration. $\sum_2$ shall be at rest, but it shall be located in a homogeneous gravitational field that imparts to all objects an acceleration $-\gamma$ in the direction of the X - axis.

As far as we know, the physical laws with respect to $\sum_1$ do not differ from those with respect to $\sum_2$; this is based on the fact that all bodies are equally accelerated in the gravitational field. At our present state of experience we have thus no reason to assume that the systems $\sum_1$ and $\sum_2$ differ from each other in any respect, and in the discussion that follows, we shall therefore **assume** the complete physical equivalence of a gravitational field and a corresponding acceleration of the reference system.

This assumption extends the principle of relativity to the uniformly accelerated translational motion of the reference system. The *heuristic value* of this assumption rests on the fact that it permits the replacement of a homogeneous gravitational field by a uniformly accelerated reference system, the latter case being to some extent accessible to theoretical treatment" (Einstein 2007, p.302; my italics).

Note that Einstein was first and foremost interested not in the ontological, metaphysical import of his principle that could enable him to elevate it up to the status of the Ultimate Law of Nature. The latter would be valid everywhere with any degree of validity being contemplated by a Super Reason trying to grasp the essences of the things and events. (For it is well-known, according to Norton 2007, that in 1907 Einstein was unaware of Eotvös's exact experimental results on the equality of inertial and gravitational mass. Furthermore, Papapetrou in 1951 had unfolded that in the GTR a rotating body falls differently, in general, from a non-rotating body).

Thus, both in STR and GTR cases he was highlighting the *heuristical* components of the principle (see Ryckman 2005 for details). In gravity purview he strove for comprehending gravitational and inertial phenomena from a single point of view. As Michele Janssen puts it,

"While the slide into general covariance turns the relativity of nonuniform motion of space-time coordinate systems into a feature general relativity shares with older theories, it does not so trivialize the relativity of the gravitational field. Even in generally covariant reformulations of these older theories, there will be an inertial field and a gravitational field existing side by side. The unification of these two fields into one inertia-gravitational field that splits differently into inertial and gravitational components in different coordinate systems is one of Einstein's central achievements with general relativity" (Janssen 2012, p.162).

In my view, it was consequent implication of the Equivalence Prin-

ciple that promised to contrive a consequence of hybrid models unifying the STR and Newton's theory of gravity. In my view the principle
of equivalence was for Einstein not so much a Law of Nature as a **pattern,** a fecund 'paradigm' for construction of empirically meaningful
gravitation theories.

In particular, it enabled to inquire into the special cases of gravitational field by means of the survey of accelerated motion. So, until
1911 Einstein had committed himself mainly to exploring, by means
of the Equivalence Principle, the effects and conceptual changes char-
acterizing a new theory of gravitation, evidently without seriously at-
tempting its construction. Only in early 1911 was he challenged by
the provoking publication of Max Abraham to elaborate such a theory,
at least for the peculiar case of a static gravitational field (see Norton
1986).

Furthermore, in September 1913 Einstein presented a lecture at
the 85th Congress of the German Natural Scientists and Physicians
in Vienna that was published in December issue of "*Physikalische
Zeitschrift*" under the humble heading "*On the present state of the
problem of gravitation*". In this fine presentation Einstein manifested
his preference for Nordström's theory over other gravitation theories.
He asserted that Nordstr?m's later version of his gravitation theory
was the only competitor to the '*Entwurf*' satisfying four constraints
that should be imposed on any reasonable theory of gravitation.

1. Satisfaction of the laws of energy and momentum conservation.

2. The equivalence principle.

3. Validity of STR.

4. The observable laws of nature do not depend on the absolute
   magnitude of the gravitational potentials.

Note that Einstein stressed the *heuristic* value of almost all the requirements admitting that "the postulates 2-4 resemble a *scientific
profession of faith more than a firm foundation*" (Einstein 1913).

On the other hand, the second substantial component of Einstein's
powerful heuristic – 'the Lorentz model of a field theory' (Renn, Sauer)

– enabled Einstein to fathom Newtonian gravitation and inertia as special cases of a more general interaction. For the case of uniform acceleration he could directly identify inertial effects with a scalar Newtonian gravitational field; and he expected that he would be able to do the same for more general cases by proper generalizing the notion of gravitational field. A paradigm for the generalization was of course furnished by relativistic electrodynamics. It was Einstein's STR that artfully amalgamated electricity and magnetism through treating electric field $\mathbf{E}$ and magnetic field $\mathbf{B}$ as different facets of one and the same electromagnetic field tensor $F_{\mu\nu}$. Accordingly, for Einstein the most important asset of the GTR was not the notorious 'geometrization of gravity' but deft 'unification of gravity and inertia' via the metric tensor $g_{\mu\nu}$.

Between 1907 and 1911 he ingeniously applied the Equivalence Principle to derive several consequences of his yet to be formulated relativistic theory of gravitation.

Note that in the case considered Einstein punctiliously follows the path of the STR. Forsooth, invention of the new theory commences with construction of the crossbred object, i.e. with the insertion of mass-energy relation into the theory of gravity. One of the important STR consequences is the tenet of equivalence of mass and energy. However, according to Einstein,

"this result suggests the question whether energy also possesses heavy (gravitational) mass. A further question suggesting itself is whether the principle of relativity is limited to nonaccelerated moving systems" (Einstein 1907, p.254).

From the very beginning Einstein was craving for such a theory of gravitation that should artfully embrace both the knowledge on gravitation and inertia represented by the classical mechanics and the knowledge on the structure of space and time layed out by the STR. However, insertion of the crossbred object– the introduction of inertial and simultaneously gravitational mass – leads to penetration of subtle STR methods into Newtonian theory of gravity and to reverse penetration of renowned Newtonian gravity methods into the STR. As a result the both theories were radically rebuilt from within and the corresponding changes in both of them were set up. The changes were epitomized in the peculiar sequences of crossbred models, the byproducts of the transformation performed.

On the one hand, an inevitable consequence of the STR penetration into Newtonian theory of gravity turned out to be Nordström's and Abraham's fine scientific research programmes. On the other hand, no less inevitable, owing to the Equivalence Principle, was the Newton theory penetration into the STR that led to the sequence of Einstein's works on the generalization of Relativity Principle and to spreading the principle not only on inertial systems of reference, but on the various accelerated systems as well. Einstein applied the Principle of Equivalence in order to transform the knowledge not of classical mechanics only but the knowledge embodied in *both*, the classical mechanics and the STR. His thought-provoking theory of the static gravitational field, as well as his early vigorous attempts to generalize it, were nothing but a reinterpretation of the STR with the help of involvement of accelerated frames of reference. His systematic treatment of such accelerated frames induced him to use generalized Gaussian coordinates in order to describe the coordinate systems adapted to these frames. It was then a natural step for him to regard the metric tensor. And with the implantation of whimsical metric tensor Einstein constructed the theoretical object that was capable of representing gravitational and inertial theoretical objects on the same footing.

By the beginning of 1912, Einstein realized that he would ultimately have to proceed beyond an uncomplicated scalar theory of gravitation. His strategy was to advance carefully in a step-by-step manner towards an ample dynamical theory. The first step in the programme consisted in scrutinizing the 'gravito-static' case, the gravitational analogue of electrostatics. However, he was already making up his mind on the second step, the 'gravito-stationary' one, the gravitational analogue of magnetostatics. His ultimate goal was to advance a theory for time-dependent gravitational fields.

In March 1911 he was able to confide to Paul Ehrenfest:

"The investigations of gravitational statics (point mechanics, electromagnetism, gravitostatics) are complete and satisfy me very much. I really believe that I have found a *part of the truth*. Now I am considering the dynamical case, again also proceeding from the more special to the more general case" (quoted from Renn 2007, p. 98; my italics).

As is well-known, in 1908-1911 Einstein had neglected gravitation, possibly because of his preoccupation with intricate problem of quanta. But this, however, is only a part of the proper explanation. The remaining part consists in that he realized how much work had to be done to arrive at an ultimate global theory able to embrace all the particular results obtained, the 'parts of the truth' as Einstein shrewdly called them, artfully transforming them into the details of a grand edifice. And, since Einstein himself had delved into the exuberance of the quanta, the problem of putting up the stout scaffolds of gravitation global theory had fallen on Abraham's and Nordström's broad shoulders.

Nevertheless, one has to keep in mind that even the pathways of advancement of these conspicuous theories were outlined by Einstein himself, especially in his ground-breaking 1907 paper. Indeed, one of renowned STR consequences states that $E = mc^2$. Since, in a gravitational field, the energy of a particle depends on the value of the gravitational potential at the position of the particle, the equivalence of energy and mass suggests that:

1. either the particle's mass m;

2. or the speed of light c (or even both) must also be a function of the gravitational potential.

These enticing possibilities, a dependence of the potential either on the speed of light c or on the inertial mass m, were later efficaciously explored by Max Abraham (1912a, 1912b, 1915) and Gunnar Nordström (1912, 1913a, 1913b) respectively.

And soon it became clear that one can easily construct such a Lorentz-invariant theory of gravitation in which the inertial and gravitational masses are equal (Nordström, 1912-1914).

Besides, Einstein's uncomplicated static gravitational theory did not provide even a hint at how the global theory should be constructed. On the contrary, a world-famous theoretician, a renowned master of classical electrodynamics, a Göttingen professor Max Abraham was one of the first scholars (along with Gustav Herglotz and Max Born) to propose that the four-dimensional line element, defining the infinitesimal distance between points in Minkowski space in terms of a constant metric, has to be replaced by a *variable* line element. Its functional dependence on the coordinates is determined by a gravitational potential associated with the variable speed of light.

It was not accidental that Einstein turned to contrivance of the global gravitational theory only *after* the publication of Abraham's first vector gravitational theory. It should be stressed that for static fields Abraham's fine theory coincides with Einstein's. But the most valuable assets of the hybrid theories of Nordström and Abraham consisted in that the both theories contained jolly *promising hints* on how the global theory could be created (Norton 1992) At first, by letting the geometry of Minkowski space *depend* on the gravitational potential (Abraham). At second, by laying out the gravitational potential not by a single function but by a ten-component theoretical object on a par with the stress-energy tensor and having this tensor as its source (Laue and Nordström). At third, by including an effect of the gravitational potential on the measurement of space and time (Nordström).

4.4 Part three. An Origin of Notorious 'Entwurf'.

Let me start from Nordström's trailblazing result doggedly derived with a help of M. Laue's fine achievements. The result draws on the exasperating problem of maxwellian electrodynamics – the problem of electron's electromagnetic mass that owes so much to Abraham's attainments (see, for instance, Abraham 1909). If one computed total momentum and energy of the electromagnetic field of an electron, the result universally accepted at that time was:

(Total field momentum) $= (4/3c^{2)}$ (Total field energy) (Electron velocity).

Hence, as Poincaré and Einstein skillfully elucidated, there must be also stresses of a non-electromagnetic character within the electron (the so-called 'Poincaré's stresses').

So, the puzzle Max von Laue addressed in 1911 was to unfold very general circumstances under which the dynamic of such an electron would agree with the relativistic dynamics of point masses. While Hermann Minkowski had cautiously introduced the four-dimensional stress-energy tensor in the course of elaboration of four-dimensional methods in STR, his use of the tensor was restricted to the special case of the electromagnetic field. Laue's immaculate work concentrated on extending the use of this tensor to the most general realm (Laue 1911a, 1911b, 1911c). The properties of the tensor and its subtle behavior under Lorentz transformations summarized a great deal of the then current knowledge of the behavior of stressed bodies.

As a result, Laue arrived at the expression for the stress-energy tensor $T_{\mu\nu}$ ($\mu,\nu = 1,2,3,4$) that embraced three important blocs.

1. The first bloc represents the common three dimensional tensor p_{ik} (i,k = 1,2,3).

2. The second bloc represents the momentum density $\mathbf{g}$ (g_x, g_y, g_z).

3. The third bloc represents the energy flux θ (θ_x, θ_y, θ_z).

And surely the (T_{44}) component of the energy-momentum tensor represents energy. Einstein's Equivalence Principle prompted that each bloc of the stress-tensor should furnish *its own peculiar impact* into the gravitational field potentials, i.e. each bloc is related to the potentials of its own. Hence there should be a throng of multifarious gravitational potentials - scalar ones, vector potentials, etc. and not a single one. That is why the overall gravitational field potential should be a group of several potentials and should in the most general case be described by a matrix, a tensor, since its components are properly transformed via the coordinate transformations like scalars, vectors, etc. The most pertinent analogy that played an important heuristic part was, of course, good old maxwellian electrodynamics with 4-dimensional electromagnetic field potential $A^\mu = (\mathbf{A}, \varphi)$. The latter, in particular static electromagnetic field case, is famously reduced to static potential φ.

It's no wonder that in the '*Zurich Notebook*', just before the '*Entwurf*' publication, Einstein had freely grappled with tensors. The traces of the work can be easily excavated in his unpublished STR review, probably written between 1911 and 1914 for *"Handbuch der Radiologie"*. The heading of the section 3, dealing with vectors, tensors, etc., speaks for itself: *"Some Concepts and Theorems of the Four-Dimensional Vector and Tensor Theories that Are Necessary for the Comprehension of Minkowski's Presentation of the Theory of Relativity"* (see Klein et al., 1992, Doc.1). Moreover, one of the subsections is promisingly entitled *"The Stress-Energy Tensor of Electromagnetic Processes"*.

This hallmark was later diligently elicited by an eminent Göttingen specialist in electrodynamics Max Abraham in his 1915 thought-provoking paper *"Recent Theories of Gravitation"* The paper contained such a substantial passage critically analyzing Einstein's and Grossmann's *'Entwurf '* that it is worth quoting in full.

"The basic idea of the tensor theory of the gravitational field can be understood as follows. The energy density, which in a static field is determined by the divergence of the gradient of the gravitational potential, plays in the theory of relativity merely the role of one component of the resulting world tensor T; it is joined by nine other tensor components which characterize the energy flux and the stresses. The tensor theory assumes that, like the energy density (T_{44}), the remaining nine components $T_{\mu\nu}$ (μ, ν = 1,2,3,4) *generate gravitational fields* whose potentials $g_{\mu\nu}$ form a ten-tensor themselves "(Abraham 1915, p.499).

The physical meaning of the components is manifested by Abraham below when he remarks that the integration of 'Entwurf's field equations

"is extraordinary difficult. Only the method of successive approximations promises success. In this one will usually take as a first approximation the solution that treats the field as static. Here, *Einstein's theory becomes identical with Abraham's theory* [...]"

In his Vienna lecture A. Einstein takes the normal values of the $g_{\mu\nu}$ as the first approximation: $g_{11} = g_{21} = g_{33} = 1$; $g_{44} = -c^2$, $g_{\mu\nu} = 0$ for for $\mu \neq \nu$; he considers the deviations $g^*_{\mu\nu}$ from these normal values as quantities of first order, and arrives, by neglecting quantities of second order, at the following differential equations:

$$\Box g*_{\mu\nu} = T^m_{\mu\nu}.$$

For incoherent motions of masses, the last (T^m_{44}) among the components of the material tensor $T^m$is the most important; it determines the potential $g*_{44} = \Phi^g$. Then follow the components T^m_{14}, T^m_{24}, T^m_{34}, which are of first order in v/c; these determine the potentials $g*_{14}$, $g*_{24}$, $g*_{34}$, which can be viewed as the components of a space vector $-(1/c)\,U^g$. The remaining components of T^m are of second order in v/c. If one neglects quantities of this order, then one only needs to consider those four potentials, and obtains for them the differential equations

$$\Box \Phi^g = c^2 \mu \qquad (60a)$$

$$\Box U^g = c^2 \mu(v/c) \quad (60b),$$

where μ is the mass density and $\Box$ is D'Alembert wave operator.

Here the analogy with electrodynamics catches one's eye. Except for the sign, the field equations (60 a, b) agree with those that must be satisfied in the theory of electrons by the 'electromagnetic potentials', the scalar one (Φ) and the vector one (A). In this approximation, *the Einstein-Grossmann tensor theory of the gravitational fields leads to the same results as the vector theory sketched in* (IA) [i.e. the theory of Abraham]" (Abraham 1915, pp. 500-501; my italics).

Abraham's solid conclusion can be re-affirmed by the contemplation of Einstein's thought-provoking paper "*On the present state of the problem of gravitation*" (Einstein 1913c).In this paper Einstein craved to demonstrate (p.1261) that the theory of 'Entwurf' reduced in suitable weak field approximation to a theory with a four-vector field potential that was formally analogous to Maxwellian electrodynamics. It was this approximation that yielded the weak field effects now labeled as 'Machian'. Just as Einstein himself had put it in the 8-th paragraph "The Newtonian Gravitational *Field*", in the weak-field approximation the Entwurf gravitational field equations take the fascinating form

$$\Box g*_{\mu\nu} = k T_{\mu\nu} \qquad (7c)$$

"We obtain the Newtonian system insofar as we introduce the following approximations:

1. Only the mass flow is regarded as the field source.

2. The influence of the velocity of the field-generating masses is neglected, and hence the field is treated as static.

3. The velocity and acceleration components in the equations of motion of a material point are treated as small quantities, and only small quantities of the lowest order are retained. Finally, we also have to assume that at infinity the $g^*_{\mu\nu}$ vanish [...]

$$\Box g = -k\rho_0 v; \quad \Box g*_{44} = kc^2\rho_0 \quad (7c')$$

The equations (7c') and (1d) show how slowly moving masses influence each other according to the new gravitation theory. To a great extent, the equations correspond to those of electrodynamics, g_{44} corresponds to the scalar potential of electrical masses up to the sign...

G corresponds to the vector potential of electric currents; the second term of the right-hand side of (1d), which corresponds to an electric field strength resulting from temporal change of the vector potential, yields precisely those induction effects, directed like acceleration, that we must expect based on our ideas regarding the inertia of energy. The vector o corresponds to the magnetic field strength (curl of the vector potential), so the last term in (1d) corresponds to the Lorentz force" (Einstein 1913c, 564).

As his correspondence and the writings testify, Einstein agreed with Nordström's assessment of the importance of Laue's work for gravitation theory. Moreover, some pieces of his 1911 and 1913 papers (his reverent proposal to call T 'Laue's scalar', for instance) evidence that he had personal contacts with Laue and discussed the intricate stress-tensor problems with him. Such personal communication is quite compatible with the fact that both Einstein and Laue were teaching in Zurich, with Einstein at the ETH and Laue at the University of Zurich (Norton 2007). It should be added that the same year Nordström also came to Zurich where supposedly he had communicated with the both researchers.

It was instantly understood that in general the '*Entwurf*' intricate equations are not covariant though; they 'remain covariant only with respect to linear orthogonal substitutions'. Yet for a long time this juncture did not deject the authors. It testifies once more that the '*Enwurf*' field equations sprung out *not* from the Covariance Principle but represented a specific *generalization* of hybrid theories of Nordström and Abraham with a help of Laue's immaculate results. (The Covariance Principle appeared to be of great importance later – when the renowned basic GTR equations were derived in 1915).

Just as Einstein alleged in his November 1913 letter to Paul Ehrenfest,

"The gravitational affair has been clarified to my *complete satisfaction* (namely the circumstance that the equations of the gr. field are covariant only with respect to linear transformations). For it can be proved that generally covariant equations that determine the field completely from the matter tensor cannot exist at all. Can there be anything more beautiful than this, that the necessary specialization follows from the conservation laws" (quoted from Klein et al, 1993, Doc. 481; my italics).

However, on the other hand, according to later reminiscences,

"The equivalence principle allows us... to introduce non-linear coordinate transformations in such a [4-dimensional] space [with pseudo-Euclidean metric]; that is, non-Cartesian ("curvilinear') coordinates. The pseudo-Euclidean metric then takes the general form :

$$ds^2 = \sum g_{ik} dx_i dx_k$$

summed over the indices i and k (from 1 to 4). These g_{ik} are then functions of the four coordinates, which according to the equivalence principle describe not only the metric but also the 'gravitational field"' (Einstein, quoted from Seelig 1955, p.55).

Certainly, the question was raised on furnishing the mathematical apparatus properly dealing with such whimsical mathematical objects. In particular, from the mathematical point of view, the task was to discover a second order differential operator for the metric tensor covariant with respect to the largest possible class of coordinate transformations. In August 1911 Einstein left Prague, where he had

taught for a year and a half, to become a full professor at the Ei-
dgenössiche Technische Hochshule (ETH). With Einstein's return to
Zurich, he started a fruitful collaboration with his old friend Marcel
Grossman. The collaboration ceased in 1914, when Einstein left for
Berlin to become a well- salaried member of the Preussiche Akademie
der Wissenschaften.

Grossmann's friendly support was needed to solve the problem.
Grossmann dug out that appropriate and exquisite mathematical apparatus had worked out at the end of the XIX - beginning of the XX-th
century by Riemann, Levi - Civita, Ricci, Christoffel et al. That is why
the first part of the '*Entwurf*' containing the gravitational field equations was written by A. Einstein, and only the second one – by M.
Grossmann (see Seelig 1955, pp. 15-16 for details).

To comprehend the interrelations between Einstein, Nordström and
Abraham's research programmes it should be stressed that *the true
commencement of the metric programme should be related to 1912*,
when in a stupendous note published on 15 February 1911 as a reply
to Einstein's critique, Abraham significantly rethought the lines with
which he had earlier referred to Minkowski's formalism. He introduced
instead the infinitesimal line element with variable metric, thus actually extending Minkowski's space-time to a more general semi-Riemann
manifold.

"In lines 16,17 of my note "On the Theory of Gravitation" an over-
sight has to be corrected which was brought to my attention by a
friendly note from Herr A. Einstein. One should read there: 'we consider dx, dy, dz and du=icdt as components of a displacement $\overline{ds}$ in
four-dimensional space. Hence $ds^2 = dx^2 + dy^2 + dz^2 - c^2dt^2$ (12)
is the square of the four-dimensional line element where the speed of
light is determined by the equation (6) [i.e. $C^2/1 - C_0{}^2/1 = \Phi - \Phi_0$]"
(Abraham 1912d, p.1056).

In this way, Abraham was the first to contrive the mathematical
representation of the gravitational potential that was to be at the
heart of the GTR, the general 4-dimensional line element involving
a variable metric tensor. Incidentally for the time being, Abraham's
alluring expression remained as isolated mathematical stunt without
profound physical meaning. Yet, on my view, it is this equation that

was referred by Einstein in his 26 March letter to his close friend Michele Besso: "At first (for 14 days!) I too was completely 'bluffed' by the beauty and simplicity of his [Abraham's] formula" (quoted from Klein et al, 1993, Doc. 377, pp. 436-437).

Driven by Abraham's stubborn persistence, Einstein in May 1911 gave up and finally acquiesced that a generalized line element $ds^2 = dx^2 + dy^2 + dz^2 - c^2 dt^2$, as suggested by Abraham's ingenious comment of three months earlier, represents the powerful key to a generally relativistic gravitation theory.

Indeed, when Einstein had accomplished the elaboration of his theory of a static gravitational field (Einstein 1912a), he took notice of the fact that his expression for the motion in a static gravitational field takes an *exactly* the same form as that given by Max Planck for a gravitation free Minkowski space, the only difference being that the speed of light is now assumed to be variable:

$$\delta \int \sqrt{-dx^2 - dy^2 - dz^2 + c^2 dt^2} = 0 \qquad (1)$$

where δ stands for the variation of the subsequent integral and c = c (x, y, z). Einstein was now completely fascinated by the possibility the equation (1) offered for a generalization beyond the special case of static gravitational fields, as was indicated by him in an eminent 'Note in proof' to his 1912a paper:

"This shows too – as has been shown for the ordinary theory of relativity by Planck – that the equations of analytical mechanics have a significant reading *far beyond* Newtonian mechanics. The Hamiltonian equation, which was the last one written down, gives an idea about how the equations of motion of the material point in a *dynamic* gravitational field are constructed" (Einstein 1912a, p.120).

Now the pertinent question became: 'what happens when one regards the motion of point particles in the presence of *general*, i.e. *non-static* gravitational fields?' – In that case the abovementioned Laue's results amended by the 'Equivalence Principle' turned out to be of special importance. And in section 1 of the '*Entwurf*' Einstein takes the abovementioned variation principle as a starting point to argue that for non-static gravitational fields, too, one should expect equation (1) to exhibit the motions of point-particles. The only difference is that now the line element on the left-hand side of the equation has to be that defined by a general metric tensor $g_{\mu\nu}$. This was the first time the metric tensor was introduced in a published paper. Three months after the '*Entwurf*", Einstein submitted a paper to the 85th conference of the German Society for Scientists and Physicians. In this paper he explicitly manifested that "A free mass point moves in a straight and uniform line according to Eq. (1), where $ds^2 = \sum g_{ik}\,dx_i dx_k$". And finally on September 1913 in Vienna Einstein delivered a lecture eliciting the physical foundations of the 'Entwurf' and those aforementioned general conditions (1) - (4) which any relativistic theory of gravity should satisfy (Einstein 1913c).

As a result, the main attainment of the second streak consisted in the invention of distinctive metric tensor; the latter turned out to be a cohesive crossbred object that deftly combined two diverse research traditions – a 'physical' tradition (scalar and vector theories of Einstein, Nordström and Abraham) and a 'mathematical' one (dainty geometrical results of Riemann, Christoffel, Levi-Civita et al.). Now the components of g_{ij} had a *dual* function: on the one hand, they laid out the physical gravitational potentials and on the other they represented the coefficients in the expression of $ds^2 = \sum g_{ij}\,dx^i\,dx^j$. By means of invention of the crossbred object g_{ij} mutual interpenetration of geometry and physics began: physics became geometrized, and geometry was made empirical (Zahar 1989, p. 267).

The asset of Einstein's scientific style of reasoning that enabled him to generalize substantially diverse hybrid models of Abraham, Nordström and his own theory of static gravitational field was his conspicuous *anti-dogmatism*. For instance, contemplating Einstein's efforts to create the special theory of relativity, Hirosige (1976) shrewdly attributed Einstein's sensitivity to the inconsistencies between mechanics and electrodynamics to influence of Ernst Mach, whose writings supposedly helped the inventor of special relativity to outdo dogmatic adherence to the mechanistic worldview. Einstein could therefore freely juxtapose Newtonian mechanics, statistical thermodynamics and Maxwellian electrodynamics without reducing one to the others (see chapter 4 for colorful details).

Jürgen Renn and Robert Schulman (1992) take Einstein's anti-dogmatism as a crucial hallmark of his distinctive scientific style of reasoning that enabled a young man to fathom the conceptual implications in the works of such boffins as Lorentz, Hertz, Poincaré and Planck that they themselves were sometimes unable to discern. Incidentally, in their Proposal for Einstein's Membership in the Prussian Academy of Science (Berlin, 11 June 1913), M. Planck, W. Nernst, H. Rubens and E. Warburg had famously professed that

"Apart from his great productivity, Einstein has a special talent for getting to the bottom of other scientists' newly emerging views and assertions, and for assessing their relationship to each other and to experience with surprising certainty"(Doc. 445 of Einstein 1987, p. 338).

Likewise, as in the STR case, Einstein was able freely to juxtapose Nordstrom, Abraham's and his own non-metric theoretical schemes without reducing one to the others.

Einstein himself, withal, did not take the geometrization of the gravitational field as a major achievement of his research program emphasizing that GTR was no more and no less 'geometrical' than Maxwell's celebrated theory of electromagnetism. For instance, in his 8 April 1926 letter to Reichenbach, he famously conveyed that

"It is wrong to think that 'geometrization' is something essential. It is only a kind of crutch [Eselsbrücke] for the finding numerical results" (quoted from Lehmkuhl 2014, p.317).

Or, likewise, in Einstein's review of Emile Meyerson's epoch-making book "*La deduction relativiste*", written in German by Einstein in 1927, it is alleged that

"The fact that the metric tensor is denoted as 'geometrical' is simply connected to the fact that this formal structure first appeared in the area of study denoted as 'geometry'. However, this is by no means a justification for denoting as 'geometry' every area of study in which this formal structure plays a role, not even for the sake of illustration one makes use of notions which one knows from geometry. Using a similar reasoning Maxwell and Hertz could have denoted the electromagnetic equations of the vacuum as 'geometrical' because the geometrical concept of a vector occurs in these equations" (quoted from Lehmkuhl 2014, p. 318).

However, one more function of 'geometrization' should be taken into consideration. I think that it was indispensable to work out a special "*neutral theoretical language*" necessary to bring the substantially different theoretical traditions – that of Newtonian gravity and Special Theory of Relativity – under the same cover to compare and to reconcile them. In this regard the following analogy with Maxwellian electrodynamics genesis seems to be especially appropriate.

In the course of electricity and magnetism true unification, the first Maxwell's stupendous paper "*On Faraday's Lines of Force*" (1856 - 1858) was dedicated to elaboration of the method of "analogies" (goaded by Kantian epistemology). The method rejects the 'ontological' opinionated approaches that strive for the dubious 'essences' of electrical and magnetic phenomena and defiantly proclaim that 'actually ' electricity and magnetism are sheer 'fields' and not 'action at a distance' phenomena, or vice versa. Maxwell's deft proposal was to take Faraday's lines of force as a kind of whimsical tubes filled with *ideal* incompressible fluid.

"I propose then, [...] ; and lastly to show how by an extension of these methods, and the introduction of another idea due to Faraday, the laws of the attractions and inductive actions of magnets and currents may be clearly conceived, without making assumptions as to the *physical nature* of electricity, or adding anything to that which has been already proved by experiment. By referring everything to the purely geometrical idea of the motion of an imaginary fluid, I hope to attain generality and precision, and to avoid the dangers arising from a premature theory professing to explain the cause of the phenomena" (Maxwell [1890], 1952, p.159; my italics).

It was crucial for a whole-hogger Kantian that the contrived incompressible poison has nothing to do with experimental reality. The constraints on the theory proposed consist in the demand that the mathematical constructs should not contradict each other. In all the other matters the method of physical analogies admits an unrestrained freedom of imagination. Even the conservation laws could be broken down.

"There is nothing self-contradictory in the conception of these sources where the fluid is created, and sinks where it is annihilated. The properties of the fluid are *at our disposal*, we have made it incompressible, and now we suppose it produced from nothing at certain points and reduced to nothing at others"(Maxwell [1890], 1952, p.162; my italics).

Maxwell punctuated the generality of the lines of force approach, for it could account for *any* kind of force. For instance, it does not exclude the force of action at a distance which varies inversely as the square of the distance, as force of gravity or as observed electric and magnetic phenomena.

And in the other parts of his masterpiece Maxwell exhibited the ways by which the idea of motion of incompressible fluid could be applied to the sciences of statical electricity, permanent magnetism, magnetism of induction, and uniform galvanic currents. The core element of his bold innovations consisted in constituting a wayward 'language game' with a 'neutral language' for description and comparison of the consequences from the rival theories. Maxwell's distinctive 'neutral language' was definitely not Carnap's and Reichenbach's notorious 'observation language' springing out of the 'protokolsatze' direct generalizations. Maxwell was well aware of the theory - laidenness of the observation data ("experimental laws already established, which have generally been expressed in the *language of other hypotheses*" (Maxwell [1890], 1952, p.162). He clearly comprehended that every observation ineradicably carries the footprints of the theoretical language that helps to describe it. ("The daubing of untempered mortar", as he colorfully called them later in his '*Helmholtz*' superlative paper – Maxwell, 1877).

In order to compare and to amalgamate in an ample theoretical scheme lacking contradictions all the results of the multifarious experiments carrying the footprints of substantially diverse theoretical languages, it is necessary to construct an *artificial* theoretical language equally distant from the wayward and queer languages of theories under comparison. This language transpired to be the language of solid state mechanics (with hydrodynamics as its part). Maxwell's ultimate aim was to rewrite all the known empirical and theoretical laws of electricity and magnetism using this neutral language and then to compare them in order to create a consistent system (Nugayev 2015).

Note that from the antediluvian 'presentation' philosophical stand-point all the hydrodynamic models were doomed to failure efforts to describe what could not be described in principle – things in themselves, the 'nature' of electrical and magnetic phenomena. On the contrary, Maxwell aimed his program to find empirically meaningful *mathematical relations* between the basic objects of electrodynamics, i.e. the creation of concordant electromagnetic field equations system.

For instance, the aim of Maxwell's second paper [Maxwell 1861-1862] was to rederive the results of the Weber and Neumann theories on the basis of a new mechanical hypothesis concerning the *vortices of incompressible fluid*.

"My object in this paper is to clear the way for speculation in this direction, by investigating the mechanical results of certain states of tension and motion in a medium, and comparing these with the observed phenomena of magnetism and electricity" (Maxwell [1861-1862/1890]1952,p.162).

Hence again and again Maxwell had to stress that

"the author of this method of representation does not attempt to explain the *origin* of the observed forces by the effects due to these strains in the elastic solid, but makes use of the mathematical analogies of the two problems to assist the imagination in the study of both" (Maxwell [1861-1862/1890], 1952,p.163).

In my view, in quite a similar vein Einstein used geometry as a neutral theoretical language in his amalgamation of gravitational and inertial phenomena (see Sauer 2015 for details). Moreover, the profound interpenetration of geometry and physics led to construction of the GTR fundamental theoretical scheme. The first stage of interpenetration resulted in the gravitational field equations of 'Entwurf:

$$R_{\mu\nu} = \chi T_{\mu\nu}$$

with their ultimately simple premises of gravitational potentials being common partial derivatives of metric. However, the further penetration of physics into the geometry led to skillful modification of the plain scheme.

"The key to this solution was the realization that not $g^{ik}d/dx\ g_{ik}$ but the related Chhristoffel symbol $\{\frac{im}{l}\}$should be taken as a natural expression for the 'components' of the gravitational field "(Einstein to Sommerfeld, 28 November 1915, quoted from van Dongen 2010, p.30).

4.5 Part four. The Masterly Contrivance of Ultimate 1915 November Mellow Theory of Gravity.

Moreover, the incessant interpenetration of geometry and physics eventually resulted in the construction of the GTR fundamental theoretical scheme. The first stage of the relentless interpenetration was accomplished by the gravitational field equations of the '*Entwurf*':

$$R_{\mu\nu} = \chi T_{\mu\nu}$$

with their ultimately simple premises of gravitational potentials being common partial derivatives of metric. However the further subtle interplay of physics and geometry led to skillful modification of the plain scheme.

In a series of four dainty publications from November 1915, submitted to the Prussian Academy of Science on the 4th, 11th, 18th and 25th, Einstein gradually replaced the vexasious '*Entwurf*' by a full-blooded metric theory of gravitation, solving incidentally the intricate problem of Mercury's perihelion. To comprehend Einstein's reasoning in passing from the '*Entwurf*' to the GTR one has to address the abovementioned Einstein's synthetic dual strategy that artfully embraced and successfully matched both a physical and mathematical approaches to finding the ultimate field equations.

The '*dual method*' suggested that Einstein 'inductively' proceeded first and foremost from a set of the following constraints of sheer *physical* nature:

(P_1) **The Newtonian limit.**

A relativistic theory of gravitation must obviously correspond to Newtonian theory in the case of weak and static gravitational fields. So, in this limit one must recover a Poisson-like equation for a scalar potential from the field equations.

(P_2) **Conservation of energy and momenta.**

On the other hand, the complementary 'top-down' (van Dongen) or 'deductive' *mathematical* approach originated from the following two requirements.

(M_1) **The Principle of Equivalence**. Observations carried through in a uniformly accelerated system are equivalent (in the abovementioned sense) to observations made in a homogeneous gravitational field.

(M_2) **Generalization of the Principle of Relativity**. The laws of physics are identical for relatively inertially moving observers to observers in accelerated motion

The (M_1+M_2) jointly imply that the field equations have to be generally covariant. Thus, a natural starting point for the mathematical approach would be to scrutinize a generally covariant mathematical abstract object that was known from the mathematics literature.

The key point is that the 'dual strategy' necessarily employed *both* physical and mathematical approaches: "it was an *iterative* process that began with trying out one approach, and then checking results with the demands or results of the other approach" (van Dongen 2011, p.11).Yet the physical approach first prevailed and lead to the '*Entwurf*'; but then the mathematics approach took over and incessantly led to 1915 November General Theory of Relativity.

Incidentally in the "*Zurich Notebook*" Einstein already possessed at his disposal the gravitational field equations that he would publish in November 1915. He had been necessarily led to them by the abovementioned 'interplay' of the dual method; however, they appeared to be inconsistent with a coherent system of 'physical' constrains. First and foremost, they did not furnish the Newton metric $g_{\mu\nu} = $ diag (g_{00}, -1.-1,-1) in the weak-field approximation. On the contrary, an obvious advantage of the '*Entwurf*" equations was that, when taking the Newtonian limit, one was not obliged to insert the notorious 'harmonic coordinates' restrictions to recover the Poisson equations in its familiar to Einstein form, and blatant inconsistencies with the Newton metric $g_{\mu\nu} = $ diag (g_{00}, -1.-1,-1) were therefore not to be expected.

Moreover, Einstein's confidence in the '*Entwurf*' was strengthened by a notorious 'hole argument' ('happenings in the gravitational field cannot be uniquely determined by generally covariant differential equations for the gravitational field').So, eventually in November 1913 Einstein was able to confess to his former assistant Ludwig Hopf:

"I am now very satisfied with the gravitation theory. The fact that the gravitational equations are not generally covariant, something that quite disturbed me for a while, is unavoidable; it can easily be shown that a theory with generally covariant field equations cannot exist if the demand is made that the field is mathematically completely determined by matter" (1 November 1913, A. Einstein to L. Hopf; quoted from van Dongen, 2010, p.22).

Yet in the light of the dual strategy and the synthetic character that a true relativistic theory of gravity should possess it is no wonder that the absence of generally covariant field equations was strongly criticized by Max Abraham (1914,p.8) and Gustav Mie (1914a; 1914b,p.176). However, initially Einstein was stubbornly blind to such criticism.

Nevertheless, A. Einstein and M. Besso applied the metric $g_{\mu\nu} =$ diag $(g_{00}, -1.-1,-1)$ to calculate for the first time the anomalous perihelion shift of Mercury. The result disappointingly appeared to be a factor of 2,4 off from the observed value.

Much distressed, Einstein would soon give up on the troubled theory; yet eventually his reasons turned out to be *threefold*. As he later confided to Arnold Sommerfeld,

"[At first] I proved that the gravitational field on a uniformly rotating system does not satisfy the gravitational field equations.

[At second] the motion of the perihelion of Mercury came out as 18" instead of 45" per century.

[At third] the covariance argument in my paper of last year does not give the Hamiltonian function H. When suitably generalized it allows an arbitrary H" (Einstein to A. Sommerfeld, 28 November 1915, Doc.153; quoted from van Dongen 2010, p.26).

As a result, Einstein found a promising way out of the '*Entwurf*' predicament: he returned to the half-forgotten mathematical strategy and to the relentless requirement of general covariance that he had harshly abandoned in the '*Zurich Notebook*'. Subsequently, in the week after 11 November 1915, Einstein resolutely altered the calculation of Mercury's perihelion shift. He realized at last that covariant equations $R_{\mu\nu} - (R/2)g_{\mu\nu} = kT_{\mu\nu}$ cannot be consistent with the Newton metric $g_{\mu\nu}$ = diag (g_{00}, -1.-1,-1).Now he was forced to recognize the substantially tensor character of the theory and had to allow the spatial components of the metric tensor to be functions of the coordinates too: $g_{\mu\nu}$ = diag $\{g_{00}(x), g_{11}(x), g_{22}(x), g_{33}(x)\}$.The ds^2 equation that Einstein applied in the November 1915 paper to account for Mercury's motion strongly resembles the Schwarzchild black hole metric with it's eminent g_{00} = (1-2m/r) and the g_{11}-g_{33} terms depending on r and m.

Digging still deeper into the mathematical background, Einstein elicited, at last, the correct value of the perihelion shift: he famously found 43", which astonishingly confirmed his desired 45"± 5" per century.

Thus, it appeared to be the reversion to the mathematical requirement of general covariance that finally brought salvation and that eventually relieved Einstein of his prejudices regarding the Newtonian limit $g_{\mu\nu}$ = diag (g_{00}, -1.-1,-1).He eloquently expressed his amazement at this truly 'enormous achievement' in a famous letter to Michele Besso:

"Read the articles! They bring the final salvation to misery. The most joyful aspect is *the accordance between the perihelion motion and general covariance*, the most striking however the circumstance that Newton's theory for the field is already incorrect at the first order of the equation (the terms g_{11}-g_{33} arise). The simplicity of Newton's theory is only due to the fact that the g_{11}-g_{33} do not arise in the first approximation to the equations of motion for a point mass" (21 December 1915, Einstein to M. Besso, quoted from van Dongen 2010, p.29; my italics).

In the light of the proposed rational reconstruction of the GTR origin the abovementioned 'accordance between the perihelion motion and general covariance' appears to be apparent. Einstein could arrive at correct explanation of the abnormal perihelion motion (grounded on the understanding that there are many different *gravitational* forces that act not only along the plane of Mercury rotation round the Sun) only when he had fully comprehended the substantially *tensor* nature of the gravitational field. That is, when he fathomed that the $g_{\mu\nu}$ is a host not only of scalar components like $g_{00}(r, m)$, but of the g_{11}, g_{22}, g_{33} components as well, each of which also properly depends on r and

m.

It is important to acknowledge the following. The ultimate GTR equations were manifested as such only in November 1915, so the whole two years had passed after the blatant discrepancy between the '*Entwurf*' and the perihelion motion was revealed in 1913. And though Einstein indeed acknowledged this discrepancy as one of his basic spurs for the 'Entwurf-GTR transition' (see the letter to Arnold Sommerfeld, 28 November 1915, Doc.153, pp. 206-209 in Schulmann et al 1998), the things were not that simple. The '*Entwurf*' really could not explain away *all* the abnormal motion. Yet Einstein quite reasonably did not take this fact as an ultimate refutation of his artful theory since the remaining unexplained parts of the precession could probably be due to other, non-gravitational factors. Einstein and Besso had found the faulty value already in June 1913, but during the two years – up to September 1915 – Einstein constantly and persistently expressed confidence in the '*Entwurf*' theory. (See, for instance, his eloquent letter to Michele Besso from 10 March 1914, Doc. 514, pp. 603-604, Klein et al. 1993). In my view, the main thrust in passing from the '*Entwurf*' to the GTR came from his abovementioned dual strategy inducing him to *match* the physical and mathematical approaches. And only *after* he efficiently reconciled them, that is, only after he conceived the general covariance of the gravitational field equations, could he turn to correct explanation of the Mercury perihelion. Hence the decisive impetus in constructing the General Theory of Relativity came from the vigorous attempts to reconcile physical and mathematical approaches, embodied in Abraham, Nordström and Einstein's deft crossbred theoretical models. Trustworthy explanation of Mercury perihelion motion appeared to be a lucky by-product of strenuous reconciliation efforts.

4.6   Part five. Significant Findings.

To recapitulate, the furnished reconstruction enables one to highlight some discernible hallmarks of the process of the GTR genesis that are commonly obfuscated by the rival approaches and to arrive at a more comprehensive account of the *Inter-Theoretical Context* of Einsteinian revolution. (One should keep in mind that real creative science is always messier and more complicated than philosophers of science and logicians like to fancy).

1. Invention of relativistic theory of gravity had dawned with the crossbred object construction in Einstein's 1907 farsighted paper. The crossbred object insertion – the discreet introduction of inertial and simultaneously gravitational mass ($E = mc^2$) – led *perforce* to STR methods penetration into Newtonian theory of gravity and to reverse penetration of Newtonian gravity methods into the STR. As a result, the both theories were radically restructured and the corresponding changes in both of them were set up. The changes were epitomized in the peculiar sequences of crossbred models, the byproducts of the transformation performed.

   (a) On the one hand, an inevitable consequence of the STR invasion into Newtonian theory of gravity turned out to be Nordström and Abraham's discreet scientific research programs.

   (b) On the other hand, no less relentless, owing to the equivalence principle, was the Newtonian theory invasion into the STR that led to the sequence of Einstein's conspicuous papers on the generalization of relativity principle and to spreading the principle not only on inertial systems of reference, but on the various accelerated systems as well. But the most valuable result of the hybrid theories of Nordström and Abraham consisted in that the both theories maintained *very promising hints* on how the global theory could be put up. Hence the climax of the stage was Einstein's daring proposal and apprehension of the fecund Equivalence Principle that became one of firm GTR heuristic foundations.

2. The basic GTR model was sedulously constructed due to the proper unification of the hybrid models of Einstein, Nordström and Abraham invented within diverse research programs. The asset of Einstein's style of reasoning was his open-mindedness,

his renowned anti-dogmatism that made possible for Einstein to freely juxtapose Nordström, Abraham's and his own non-metric theoretical schemes without blunt reducing one to other.

3. It is this asset that helps to comprehend the true reasons for Einstein's victory over the rival programs of Nordström and Abraham. The 'Entwurf' and the GTR artfully 'incorporated' Nordstrom's and Abraham's deft accounts in a higher level theory. Thus, a true reason for Einstein's victory over the rival programmes does not consist in that he peremptorily rebutted the alternative accounts. On the contrary, it consists in that he subsumed them all into a new mellow and grand framework.

Hence the decisive drive in advancing the General Theory of Relativity came from the craving attempts to match physical and mathematical approaches, embodied in the Abraham, Nordström and Einstein's crossbred theoretical models. Trustworthy explanation of Mercury perihelion motion turned out to be a natural by-product of strenuous reconciliation efforts.

4. The monumental volumes (Renn et al., 2007), which were repeatedly cited in the present account, are permeated with the idea that GTR "was not just a one-man affair and a lucky individual discovery but the result of a profound transformation of the extended system of knowledge of classical physics". According to the authors, the genesis of GTR can be taken as an example of sheer 'Copernican process' that is repeatedly observable at decisive moments in the history of science.

"This process implies that marginal elements of a complex architecture of Knowledge torn asunder by inner tensions become the starting points for a reconstruction as a result of which a new structure emerges that is, however, essentially composed of the building blocks already available. This process may be designated as a "Copernican process". Indeed conceptual turnovers unfold in a similar way to the revolution of Copernicus, who created a new world system by placing an initially marginal celestial body – the sun – into the center, while, instead of starting with a *tabula rasa*, retaining the previously elaborated complex machinery of planetary astronomy" (Renn and Sauer 2007, p.47).

In a sense this chapter can be taken as an attempt to *explicate* and *furnish* the aforementioned doctrine of the 'Copernicus process', rendering a *mechanism* of simultaneous profound transformation of the classical paradigms. The hallmark of the mechanism is the construction of the crossbred systems from the basic theoretical objects of 'old' mature theories (chapter II; see also Nugayev 1999, 2018). These crossbred systems are subsequently generalized to form a structure of a novel mellow theory. A mellow physical theory can be empirically successful if it represents a kind of an 'entwurf' that can be explicated by the results of new experiments that increase its empirical content. In this way GTR could become such a vehicle for production of new gravitational propositions since it managed to reconcile and intertwine the 'old' classical traditions of Newtonian mechanics (including Newtonian gravity) and special theory of relativity in such a way that they ceased to contradict each other and their newly constructed fusion could successfully explain and predict new experimental data.

Finally, the focus on history of GTR broaches the following subject: what can one learn from Einstein's contrivance of startling 1915 ne plus ultra theory about how science actually works?

In my judgement, the abovementioned story can better elicit the subtle interconnections between the two common accounts of the scientific method in the GTR that are tightly connected with renowned empiricist and rationalist epistemic traditions.

According to the first one (see, for instance, Popper 1963; Lakatos 1970; Kuhn 1961), theories, research programs and paradigms are concocted by theorists to comprehend the phenomena which experimentalists have previously sedulously discovered. When a gallant 'paradigm' clearly accounts for all the available experimental data, it is applauded as an 'incredible success'. Yet sometimes a novel experimental phenomenon is elicited which was not predicted by the paradigm, making it 'anomalous'. From now on, a new conspicuous paradigm is artfully invented (or a new solid 'research program' is conjured up) to explain the exasperating anomaly away. Incidentally the promising innovation must account for all the 'old' observations and experiments also. When the 'new' paradigm renders a correct description of *all* the experiments at hand, it is considered as a proper 'replacement' for the decrepit one. Yet the hard problem with this blunt and lucid empiricist epistemic model consists in that it does *not* apply to the history of general theory of relativity. Almost all the sheer general-relativistic phenomena were unknown in 1915 when Einstein had arrived at his startling theory. These include the gentleman's set embracing the Multiverse, the expanding universes, black holes, naked singularities, gravitational waves and so on. Not only were these queer phenomena not yet fixed in 1915, most of them had not even been imagined.

Just compare the odd problem situation with proverbial history of quantum theory. It has to be acknowledged that a throng of 'characteristic quantum phenomena' (Smolin 2015) were well-known to physicists before quantum mechanics was put up in 1925-26.

Empiricists commonly show off the eminent Mercury's perihelion shift as a textbook kind of the anomaly that GTR had successfully explained away. But the stout problem with the perihelion consists in that only Einstein and a small circle of advanced theoreticians connected with him conceded that the true comprehension of this phenomenon needed a *radical breakthrough* with the classical research traditions. The phenomenon was well-known and thoroughly discussed half a century before Einstein, and the majority of astronomers had maintained that the perihelion shift could be duly explained either by a new planet or by more reliable and fine calculations based on standard and entrenched classical premises.

Moreover, the blatant discrepancy between the '*Entwurf*' and the perihelion motion anomaly was found *already* in 1913. But Einstein quite reasonably did not regard this fact as an *ultimate refutation* of his artful theory since, as was abovementioned, the remaining unexplained parts of the precession could be probably due to other, nongravitational factors. So, the actual things were not that simple as the empiricists maintain. Mercury's perihelion motion was not judiciously taken as a murderous evidence against Newton's theory of gravitation by the bulk of the physics community. In contrast, what should have been quite clear to *any* physicist was that Newton's theory of gravitation blatantly contradicted Einstein's special theory of relativity. *It was this blatant and commonplace anomaly that ought to be first and foremost explained away after 1905.*

But why not concoct a field theory for gravity within the perspicuous framework of special theory of relativity? – This was the prudent route Einstein and several of his contemporaries including Abraham, Nordström et al. took; some of them continued to follow this path even until Eddington's 1919 crucial observations of light bending by the Sun. Thus, the pertinent question is why did Einstein doggedly ignore this beaten track in favor of dubious and risky game?

And here comes another renowned myth – now a rationalist one. In this instance Einstein is depicted not as Humean and Machian confounded empiricist which 'ferrets out' (Einstein's own words) the 'first principles' but as a Byronic rebellious spirit, a 'lone genius' who applies exquisite mathematics to put up the sublime edifice of his grand theory. Note that no one was more responsible for inventing this myth than Einstein himself (see, for instance, his renowned 1949 'Autobiography', pp. 88-89). The common trouble with the second myth is that it is at odds with history of science; we have, quite luckily, the 'Zurich Notebook' for deprecation

The point of view presented in this chapter constitutes a distinctive attempt to find our own wayward and 'coherentist' way between the Scylla of down-to-earth empiricism and the Charibdus of sublime rationalism by showing off the reconstruction that contains both valuable epistemic insights as simple momenta of a more intricate account. The exasperating problem situation that led Einstein to GTR was engendered not by the Mercury perihelion anomaly *per se* but by the blatant and substantial discord, the profound cross-contradiction between the new fangled special theory of relativity and Newton's celebrated theory of gravitation. Yet to resolve it, i.e. to sedulously contrive a global theory, Einstein had to take into consideration all the significant results obtained in hybrid theories of Abraham, Nordström and of himself. The crux of his pre-eminent theory – the fundamental theoretical scheme of GTR – is a specific generalization of the crossbred theoretical schemes of Abraham, Nordström and Einstein on the basis of the celebrated 'Principle of Equivalence'. The latter professes the equivalence of gravitational and inertial mass and hence manifests that there should not be a glaring chasm between gravity and inertia. Einstein's immaculate theoretical scheme was better than the deft schemes of Abraham and Nordström not because it was the 'true one' and its rivals were bluntly and obviously 'wry' but because it embraced them all though in rather modified forms.

Accordingly, the core of Einstein's *modus operandi* was set up by the so-called '*dual strategy*' that matched both a physical and mathematical approaches to finding the ultimate field equations. In inventing his celebrated equations Einstein was impelled to take into account the results of proverbial observations and the well-known experiments but for him they represented first and foremost the indispensable means for checking the efficacy of his coordinationist efforts "the most joyful aspect is *the accordance* between the perihelion motion and general covariance" (Einstein, 1915, abovementioned letter to Michele Besso).

The fine idea that a new theory can be contrived to challenge the entrenched paradigm, even in the sheer absence of experimental anomalies, was famously furnished by Paul Feyerabend (see, for instance, Feyerabend 1975). He alleged that this strategy can succeed when the alternative paradigm suggests new interpretations of already known experimental results and when these new interpretations can provide the performance of new experiments that distinguish it from the dotard paradigm. In this sense the present account can be taken as such an elucidation of Feyerabendian promising ideas that craves to explicate the necessary conditions under which the invention of such an alternative is possible. To sum up: the GTR was able to anticipate an avalanche of observational data concerning modern inflationary cosmology and relativistic astrophysics since it managed to reconcile two leading research traditions of the XXth century beginning that of whimsical special theory of relativity and Newton's celebrated theory of gravitation.

5 ADVANCEMENT OF THE ELECTROWEAK THEORY: A MODERN CASE STUDY

From time immemorial, man has desired to comprehend the complexity of nature in terms of as few elementary concepts as possible. Among his quests – in Feynman's words – has been the one for 'wheels within wheels' – the task of natural philosophy being to discover the innermost wheels if any such exist. A second quest has concerned itself with the fundamental forces, which make the wheels go round and enmesh with one another. The greatness of gauge ideas – of gauge field theories – is that they reduce these two quests to just one; elementary particles (described by relativistic quantum fields) are representations of certain charge operators, corresponding to gravitational mass, spin, flavor, color, electric charge, and the like, while the fundamental forces are the forces of attraction or repulsion between these same charges. A third quest seeks for unification between the charges (and thus of the forces) by searching for a single entity, of which the various charges are components in the sense that they can be transformed one into the other.

Abdus Salam

Abstract. In light of the lucid epistemic model of reconciliation and synthesis of theories approbated on the history of science material related to the genesis of Maxwellian electrodynamics and Einstein's special and general theories of relativity, the process of unification of

194

theories of electromagnetic and weak interactions that took place in the second half of the XX century is scrutinized. It is manifested that the epistemic specificity of the episode in the history of science is primarily expressed in the fact that the unification was carried out as a part of the substantially synthetic program that aimed to put up a sequence of intricate theories of electroweak interactions as consecutive approximations to the Yang-Mills gauge theory. The milestones of the implementation of the synthetic gauge program that culminated in the advancement of the so-called 'Glashow-Weinberg-Salam model', the dogged proof of its renormalizability and empirical adequacy are examined. A renowned tenet according to which the electroweak theory itself does not provide a full account of how the fields are mixed by furnishing direct values for the Weinberg angle, i.e. that the electroweak theory does not determine in any strict sense that the fields *must* be mixed is vehemently disputed. However, it is contended that in many important respects the hybrid nature of this model engenders its unfinished nature.

Keywords: electroweak theory, synthesis, epistemology, gauge theory.

5.1 Introduction. The Electroweak Theory Genesis as a Case Study.

As is well known, one of the important (if not the most one) aims of modern physics is the creation of the enticing 'Grand Theory of Everything', which can finally contemplate electromagnetic, gravitational, strong and weak interactions from a single point of view (see e.g. Greiner and Muller, 2009). In this direction, in the second half of the XX century, indubitable progress was achieved, first connected with the working out of the so-called 'Standard Model' of elementary particles and the peremptory 'Grand Unification' theories of all the fundamental interactions, except gravitational. Moreover, there were accruing hopes that the rapidly expanding inextricable theory of superstrings would finally unite all the four basic forces of Nature.

Nevertheless, the unrestrained proliferation of superstring models stretched over many years, their excessive number and extremely weak connection with the experiment impel one to look backward, returning to the origins - the first renowned 'Glashow-Weinberg-Salam model', successfully amalgamating electromagnetic and weak interactions, and to soberly evaluate both its multifarious advantages and pitiful shortcomings.

On the other hand, the asset to concrete reconstructions of humble episodes that are relatively far from the first and second scientific revolutions, which is characteristic of modern, post-Kuhnian philosophy & history of science, provokes one to sift the modern material to test the often unsteady, fragile and abstract epistemic constructions.

Therefore, *this chapter* aims to posit a novel rational reconstruction of one of the most bright episodes of post-war physics - the advancement of the electroweak theory – examined in the light of the lucid epistemic model of mellow theory change contrived in chapter I. This reconstruction is designed first to check the ingenuous epistemic model, yet also to unfold the specificity of the electroweak theory genesis, as well as the distinction of this juncture from the classical ('paradigm') synthesis samples inextricably connected with the names of Newton, Maxwell, and Einstein.

5.2 Part one. A Tough Way to the Glashow-Weinberg-Salam Deft Model.

As is well known, the first theory of weak interactions was advanced by Enrico Fermi to explain the inextricable experimental data concerning the β-decay process.

"A β decay is the process by which a nuclear neutron changes into a proton at the same time as an electron, which is observed as a β ray, and neutrinos are emitted by the described mechanism" (Fermi [1934], 1968, p.1154).

The sedulous examination of β - decay represented one of the main trends of studies of radioactivity, and up to the 30-th, no one could explain the continuity of the spectrum of emitted electrons. However, Wolfgang Pauli rendered a powerful key to fathom this phenomenon: β – decay involves two particles, an electron, and a neutral lepton, called a neutrino. Later, Fermi put on Pauli's abstract hypothesis 'in flesh and blood', contriving on its basis a quantum field theory of the force responsible for β — decay — the odd force of weak interactions. Moreover, it is crucially important for the present chapter *that the Fermi theory turned out to be the place of thrilling encounter of two substantially diverse mellow theories* - the entrenched theory of electromagnetism and the new-fangled theory of weak interactions.

"Nevertheless, to understand that β-emission is possible, we want to try to construct a theory of the emission of lightweight particles from the nucleus *in analogy* with the theory of emission of light quanta from an excited atom by the usual radiation process" (Fermi [1934], 1968, p. 1151).

Subsequently, the Fermi theory was considerably modified to successfully explain the processes of nonconservation of parity, and relentlessly turned into the V-A theory (Feynman and Gell-Mann, 1958). Nevertheless, an encounter of theories of such substantially diverse interactions as weak and electromagnetic (just compare the extremely short radius of action of a weak force - 10^{-15} - with the actual infinity of electromagnetic interaction r^{-2}) unavoidably lead to the exasperating non-renormalizability paradox and, accordingly, to a weird prediction, according to which the cross-sections for neutrino scattering by electrons should grow linearly with increasing energies. (It should lead to bizarre excess of neutrinos in bubble chambers and cosmic rays). These queer paradoxes too well resembled the notorious 'ultraviolet catastrophe' engendered by the encounter of Maxwellian electrodynamics, Boltzmann's statistical mechanics, and classical thermodynamics (see chapters II-III).

Owing to strenuous attempts to overcome the flagrant flaws of Fermi and Feynman & Gell-Mann pioneering research, such an explanation of the phenomenon of radioactivity was offered that was no longer grounded on the shabby ideas on the direct point interactions of weak currents, but on the splendid concept of α-*particle exchange*. As Abdus Salam aptly recollected later,

"By the same token, the so-called weak nuclear force - the force responsible for β-radioactivity (and described then by Fermi's non-renormalizable theory) had to be mediated by some unknown zero-spin mesons *if it was to be renormalizable*" (Salam 1984c, p.325; my italics).

To preserve the spatiotemporal structure of the V-A theory, it was necessary to assume that the carrier particles of queer weak interaction are spin 1 vector mesons, which are now called intermediate vector bosons (IVBs). Incidentally, the intricate phenomenology of V-A theory required the introduction of not only IVBs but W^+ and W^- (Morrison 2007). However, paradoxically, a change in the way of describing interactions led not so much to the resolution of the above paradoxes (i.e. to salving the 'cross contradictions' in our distinctive vocabulary), but to an even greater extent highlighted the abyss separating photons and W- bosons. The first striking difference consisted in the fact that the photons themselves are electrically neutral, while β-decay necessarily included the exchange of electrically charged particles.

However, the second - and this time basic - difference consisted in that the W bosons were *massive* particles (as opposed to a massless photon), which was quite necessary to explain the extremely small radius of action of weak forces in atoms. This hallmark (coupled with the vector nature of the intermediate vector bosons) led to the appearance of bothering divergent terms in propagators (Green's functions) for weak interactions that could not be withdrawn using common recipes borrowed from quantum electrodynamics (Morrison 2007).

Nevertheless, failures in making away the divergences manifested a substantial circumstance: it is necessary to proceed from pathetic attempts to eliminate particularities to stout salving the pivotal problem - the craving synthesis of theories of electromagnetic and weak interactions. Besides, the first *real* attempt to put up such a theory was the work of Julian Schwinger (1957), which indicated the main direction of successful unification - the gauge theory of Yang and Mills.

In a sequence of trailblazing studies, but first and foremost in their epoch-making 1954 paper Yang and Mills had enunciated that specific, coordinate-dependent, local so-called '*gauge*' transformations do not change the Lagrangian in quantum electrodynamics (and, therefore, all the predictions of this theory). Incidentally, the part of the Lagrangian that relates to electrons and positrons is not invariant under gauge transformations; invariance requires that part of the Lagrangian, which describes the *interactions* of photons with field sources. Thus, the very existence of photons interacting with field sources is a necessary condition for gauge invariance of the entire electromagnetic Lagrangian.

Whereupon, the very existence of photons interacting with field sources turns out to be a necessary condition for gauge invariance of the *entire* electromagnetic Lagrangian.

Accordingly, Yang and Mills had arrived at the farsighted idea that any field theory (and even the theory of strong interactions) can be modeled following quantum electrodynamics. In the trenchant language of group theory, gauge transformations that leave the Lagrangian of quantum electrodynamics invariant belong to the U (1) unitary group of transformations. Yang and Mills sought to contrive a similar theory that would be invariant under local transformations of the group of strongly interacting isospin symmetries - a special unitary group SU (2). The lead of the first two members of the Lagrangian, which in QED (quantum electrodynamics) related to electrons and protons, was now performed by members related to the isospin multiplets of hadrons. Now the wave function displayed a two-component field (p, n), representing a doublet of nuclear isospin. As in QED, this part of the Lagrangian turned out to be non-invariant concerning gauge transformations. However, if one discreetly introduces W-particles with spin 1 that make up a triplet (W^-, W^0, W^+), then he can, without much effort, directly insert such parts describing interactions of W-bosons with nucleons in the Lagrangian that its invariance regarding gauge transformations will be provided. As a result, a jolly odd theory advanced by Yang and Mills began to be taken as a model (a 'paradigm') of a whole class of gauge theories, embracing even the theory of gravity. This was primarily due to the astonishing fact that

"a remarkable property of non-Abelian gauge symmetry is that it not only imposes a restriction on particle masses and coupling constants but also *determines the dynamics of the interaction* of gauge fields" (Okun 1990, p. 185; my italics).

Young and Mills had contrived the SU (2) - invariant theory of interacting nucleons, and numerous groups of researchers in concert extended their results to other gauge groups, to the interactions of other particles. As T. Cao rightly notes,

"Methodologically, it was gauge theory driven by the desire of having a universal *principle* to fix a unique form of couplings among many possibilities" (Cao 1997, p.269).

But let us return to the dizzying research program of Julian Schwinger, who set as his ultimate goal to arrange the *entire* set of elementary particles in the alluring framework of the theory of quantized fields; in particular, to regard massive charged vector bosons and massless photons as gauge mesons.

Largely, his seminal contributions consisted in discreet substantiating the following two points:

1. that leptons carry a weak form of isospin similar to strongly interacting particles;

2. that the photon and the intermediate vector boson (identified as a Z-particle) constitute inextricable parts of the same isospin multiplet.

The pivotal part of Schwinger's distinctive work was the farsighted identification of the lead that symmetry played both in the contrivance of field-theoretical models of elementary particles and the subsequent synthesis of these models. He daringly suggested that one of the reasons for the diversity of elementary particles consisted in the presence of the so-called *'internal'* degrees of freedom along with such stale characteristics as a half-integer spin for fermions and a whole spin for bosons.

The hard core of Schwinger's research program consisted of the basic principles of symmetry and field theory, with the help of which he solidly worked out an immaculate conceptual framework for arranging the fundamental interactions. In particular, the leptons of weak interactions began to be taken by him as representations of a three-dimensional group of rotations from which their main features could be read. Incidentally, grounded on a consideration of the symmetry between the isospin properties of heavy bosons and fermions, Schwinger posed the enticing question of the existence of a family of bosons that would be a realization of a three-dimensional group of rotations. Because the electromagnetic field was supposedly the third component of the three-dimensional isospin vector, it was quite natural to surmise that two more charged particles enter the triplet.

Thereafter Schwinger professed that

"We are forced to accept the concept of the existence of a boson family with spin equal to unity, which includes a massless neutral photon and a pair of electrically charged Z-particles, which supposedly carry mass by analogy with leptons" (Schwinger 1957, p.434)

Also, based on the general assumption that there exists a family of bosons that is an isotopic analog of leptons, on a par with the identification of a photon as a neutral member of this family, Schwinger sedulously advanced a dynamic theory of a charged field of Z particles with spin 1.

As it often happens, the first actual attempt to combine weak and electromagnetic interactions engendered a whole bunch of problems; the main conundrum turned out to be the inextricable problem of the vector boson mass. The radius of action of weak interactions is extremely small, which, by the uncertainty principle, implies jolly significant masses of W-particles that one must insert into the Lagrangian 'by hands'. Yet this, in turn, is at odds with one of the necessary conditions for the applicability of the gauge approach, according to which the gauge particles should be *massless*.

The 'mass problem' became a byword and has led to the consensus that only partial symmetries, i.e. the invariance of only part of the Lagrangian concerning the group of infinitesimal transformations, can inextricably link the massive bosons with a massless photon. This forced the researchers to go beyond the triplet and insert an additional neutral boson Z^0, which could be associated with its neutral lepton current (t' Hooft 1980).

The next important step in the unification of electromagnetic and weak interactions was taken by Sheldon Glashow, Schwinger's star graduate student at Princeton University. In 1961 he put forward a daring theory that anticipated many features of future peremptory advances. His enticing model artfully embraced both a triplet and a singlet of gauge vector bosons, and the corresponding gauge symmetry now belonged to the SU (2) x U (1) group. The triplet contained a positively charged, negatively charged, and neutral vector bosons, while a singlet belonged only to the neutral one. Due to the corresponding amendment of the quantum-field Lagrangian, which consisted in the skillful introduction of 'mass' terms, Glashow exhibited that the singlet and the neutral terms of the triplet 'mix' so as to produce both a very massive particle Z^0 and a massless particle that closely resembles a photon (see Pickering 1984 for colorful details). Similar important results were promptly obtained - this time in the UK - by Abdus Salam and James Ward (Salam and Ward, 1964). Their punctilious paper, although it suffered from an incontestable drawback — the stale 'mass problem' of vector bosons, which were introduced into the Lagrangian by 'hands' and led inevitably to the non-renormalizability of the theory, - indicated the opportunities of resolving the inextricable problem situation. These enticing possibilities turned out to be famously connected with the concepts of '*spontaneous symmetry breaking*' and the '*Higgs boson*'.

It is crucially important, for the present chapter, that the whimsical concept of spontaneous symmetry breaking was intruded into elementary particle physics from a jolly distant field of research - solid-state physics (superconductivity). The necessary cohesive link of this trans-

fer was scientists – major authorities in two distinct fields simultaneously - solid-state physics and elementary particle physics. The lead
was played by researchers who had a solid experience *both* in group
theory and in the physics of strong and weak interactions, just as in
the case of developing gauge theories.

The very splendid idea of spontaneous symmetry breaking sprung
out of solid-state physics with the theoretical reproduction of the phenomenon of *ferromagnetism*. A common magnet is a collection of
spinning particles located in iron atoms. The very phenomenon of
magnetism owes its existence to the wayward interaction of spins when
each spin behaves like a tiny magnet. Further, the Lagrangian of the
system as a whole does not contain a hint of any particular direction,
i.e. it is invariant concerning rotations. However, when the spins of
the entire system unanimously line up to produce the phenomenon
of ferromagnetism, the physical state of the ferromagnet loses the invariance inherent in the Lagrangian. A similar phenomenon is also
characteristic of the queer phenomenon of superconductivity. From
the history & philosophy of science discreet standpoint, the situation
is analogous to the discovery of 'electromagnetic waves' in classical
electrodynamics (see below).

This is the phenomenon of spontaneous symmetry breaking, the
striking importance of which for strong and weak interactions was comprehended by Japanese physicist Yohiro Nambu. Whereupon, the title
of the first of two daring papers published in 1961 by Y. Nambu and
G. Iona-Lazilio looks like this: *'A dynamic model of elementary particles based on the analogy with superconductivity'*. The most promising
part of these studies was the daring transfer of the ideas of spontaneous
symmetry breaking into elementary particle physics: in field theory,
situations are possible where the symmetry of the system is 'hidden'.
Outwardly, the symmetry is broken, but the Lagrangian remains stubbornly invariant.

Further development of this enticing idea was associated with studies from the 60s (Higgs 1964, 1966). The so-called 'Higgs mechanism'
was proposed grounded on the perspicuous model, artfully borrowed
from QED. It consisted of a conventional electromagnetic Lagrangian,
supplemented by parts related to two scalar fields with zero spins,

which managed to interact both with photons and with each other so that the gauge invariance was preserved. Higgs manifested that if the term in the Lagrangian related to the masses of scalar fields is negative, then the physical spectrum of the theory will have to contain one massive photon and one massive scalar particle - the celebrated 'Higgs boson'. The physical import of the queer mechanism consisted in that massless photons can be known to be polarized in only two directions, while massive vector bosons have three possible directions of polarization. Therefore, a massless photon can be taken as 'eating' a scalar particle, which provided the third direction of polarization and at the same time bestow photon with mass.

Finally, *a decisive step* in the unification of electromagnetic and weak interactions was taken in 1967 independently by Stephen Weinberg (Massachusetts Institute of Technology) and Abdus Salam (Imperial College, London). Both Weinberg and Salam proceeded from the models of electroweak interactions previously proposed by Glashow (1961) and Salam and Ward (1964), but they prudently replaced terms in the Lagrangian that refer to the masses of intermediate vector bosons and introduced 'by hands' with those terms that were engendered by the Higgs mechanism. In particular, in Weinberg's deft work, the relations between the masses of intermediate vector bosons were determined using the notorious Weinberg angle θ_w (Weinberg 1967).

As Salam and Ward picturesquely noted in the introduction to their epoch-making paper,

"our basic postulate is that it should be possible to *generate* the terms of strong, weak, and electromagnetic interactions with all their correct symmetry properties (as well as with keys considering their relative forces) due to *local gauge transformations* of the terms in the Lagrangian related to the kinetic energy of all particles. This is a statement describing an *ideal* that, at least in this article, is only partially realized" (Salam and Ward 1984, p.331; my italics).

As a matter of fact, both Weinberg and Salam without further ado combined the Higgs mechanism with early electroweak models, which explains the fact that their work initially almost did not arouse significant interest in the scientific community (for more colorful details, see Pickering 1980). However, everything changed ever since impeccable proving the renormalizability of the Weinberg-Salam model in the stout writings of Weltmann and t' Hooft. As Sydney Coleman wittily remarked, t' Hooft's work turned the Weinberg-Salam frog into a charming princess.

Finally, we came to the main point of the chapter: without a doubt, the unification of electromagnetic and weak interactions took place. However, what is behind all this? In what sense should the synthesis be comprehended? What is its specificity and difference from four classical 'paradigmatic' syntheses - the Newtonian unification of 'the physics of the Earth' and 'the mathematics of the Heaven', the Maxwellian celebrated unification of electricity, magnetism, and optics, and, finally, the Einsteinian pre-eminent STR and GTR, artfully combining 'space and time', 'motion and matter'? In my judgment, the following Salam's words uttered in Stockholm during the receipt, together with Sheldon Glashow and Steven Weinberg, of the 1979 Nobel Prize in Physics can serve as the key to conceiving this specificity (see the epigraph)

"A third quest seeks for *unification* between the charges (and thus of the forces) by searching for a single entity, of which the various charges are components *in the sense that they can be transformed one into the other*" (Salam 1984c, p.324; my italics).

5.3 Part two. Margaret Morrison on the Electroweak Unification: Pro et Contra.

In my view, the starting point for the inquiry into the assets of the Weinberg-Salam model is a thought-provoking book by Canadian historian and philosopher of science Margaret Morrison (2007), which renders the most comprehensive rational reconstruction of the episode presented so far. Nevertheless, the circumstances of the origin of the Weinberg-Salam model, revealed by another epistemic model, do not allow the author of the present monograph to agree with *all* its corollaries (this will be discussed in more detail below).

The main results of M. Morrison, relevant to the topic of this chapter, are as follows.

1. The case of unification of electromagnetic and weak interactions belongs to the synthetic, but not the reductionist types of synthesis of theories (for more details on these two types see chapter II and Nugayev 1999). Thus, the unification of these interactions did not occur as a result of blunt reduction of one interaction to another, but due to the *deft combination* of A_μ and W_μ gauge fields.

2. Despite the evident synthetic method of adjustment, particles that transport interactions remain distinct from each other. However, the unification method, which is grounded on the gauge group SU (2) x U (1), unavoidably leads to ineradicable mixing of electromagnetic and weak fields. This inextricable mixing became possible due to the identification of leptons with the isospin symmetry group SU (2), which was then combined with the group U (1).

3. The main parameter describing the mixing process is the so-called '*Weinberg angle*', which is a whimsical combination of coupling constants of electromagnetic and weak interactions. Forsooth, using gauge-theoretical constraints, one can theoretically reproduce the dynamics of an electroweak model from the immaculate mathematical framework of gauge theory. However, this stout framework tells us nothing about the intricate mixing of the two fundamental fields.

4. Two mellow theories, each governed by independent symmetry groups, 'are combined under a larger symmetry structure yielding a new hybrid theory' (Morrison 2007, p.109). Nevertheless, we are dealing not with simple conjunction, but with a *genuine* synthesis, leading both to a rethinking of the electromagnetic potential and a new dynamics of both fundamental fields.

"Two theories, each governed by independent symmetry groups, are combined under a larger symmetry structure, yielding a new *hybrid* theory" (Morrison 2007, p.109).

5. The process of deft combining electromagnetic and weak forces was determined by queer considerations rooted more in the immaculate mathematics *of gauge theory* than in the inextricable phenomenology of *physical* processes. Therefore, the intricate mechanisms involved in the unification process tell us more about the multifarious varieties of subtle mathematical models and structures at the disposal of theorists than about the *ontological status* of the electroweak processes

themselves. The interaction of electromagnetic and weak fields, for example, is not described at all due to the involvement of the phenomenological facets of the electroweak theory. The blunt insertion into the 'living tissue' of the theory of a foreign element borrowed from the remoted theory of superconductivity of liquid helium - the odd concept of 'spontaneous symmetry breaking' - allows one to adapt, coordinate, and reconcile substantially diverse types of particles concerning to the electroweak theory. The properties of these particles are not only and not so much different as *diametrically opposite* (i.e. massive and massless bosons). These incompatibilities were artfully resolved using the renowned 'Higgs field'. Moreover, the symmetry groups U(1) and SU (2), governing electromagnetic and weak interactions, were artfully combined into a large group U (1) x SU (2), in contrast to the reduction of particles and forces to a common source. Nevertheless, the craved unity was achieved not due to the primitive conjunction of two theories or two groups of symmetries, but due to the introduction of *new components* into the encountered theories of weak and electromagnetic interactions.

6. Not only the unity but also the *renormalizability* of the electroweak theory owe their ability to those structural features that were imposed not so much by the phenomenology of electromagnetic and weak interactions as by the immaculate *mathematical framework* of the Weinberg-Salam model.

7. Although the electroweak theory does lay off the behavior of two fundamental fields,

"No *causal* account is given as to *why* the fields must be unified. That fact emerges as a consequence of classifying leptons as weak isospin doublets. The Higgs mechanism is a necessary feature of this unity in that sense that it allows for its *possibility*; nevertheless, the actual unification comes from constraints imposed by the isospin symmetry group and the non-Abelian structure of the fields. Although the unity is represented in the theory by the combination of the weak and electromagnetic coupling constants in the Weinberg angle, the idea of a unifying formal structure is crucial for understanding not only how the unity was produced, but also how its nature should be understood "(Morrison 2007, p.135; my italics).

8. The subtle mechanism of combining the two mellow theories consists in the representation of the *interaction* or '*mixing*' of various fields. Yet, because the fields remain diverse from each other, the electroweak theory preserves two different coupling constants: the constant q associated with the electromagnetic field U (1), and the constant g for the gauge field SU (2). For the theory to predict the masses of the W^+, W^- and Z^0 particles supposed in this case, one must know the magnitude of the ground state of the Higgs field $/ \phi_0/$. Regrettably, this value cannot be calculated.

To get out of the trouble, the coupling constants were aptly combined into a single parameter, which became known as the Weinberg angle θ_w.

More precisely,

$$\sin \theta_w = q/\sqrt{g^2 + q^2}.$$

To determine θ_w, one must know g and q; but directly they are immeasurable. As a result, the pivotal angle θ_w , *which determines the degree of mixing of the electromagnetic and weak fields, is not determined from the theory*! This parameter is determined phenomenologically, from an experiment, and is approximately equal to 30^0.

9. The substantial empirical predictions of the electroweak theory include both the existence of W^+, W^- and Z^0 bosons and the existence of the so-called 'neutral currents'. Both predictions were confirmed (in 1973 and 1983).

In my judgment, the finest asset of Morrison's rational reconstruction is a thorough analysis of the *mathematical elements* of the unification mechanism, which elicits a significant separation of the electroweak theory from the stout empirical basis, caused by the application of concepts and methods developed in areas jolly far from distinctive physics of weak interactions. Grounded on her fine analysis, Margaret Morrison makes an emphasis on skepticism:

"What I want to show in the closing section is how the synthetic unity displayed by the electroweak theory is insufficient for drawing any conclusions about unity in nature" (Morrison 2007, p.135).

One of the arguments in favor of this distinctive standpoint is the conclusion that

"But like displacement in Maxwell's theory, a good deal of the explanatory power of the electroweak theory comes from a mechanism (Higgs) for which there is no experimental evidence" (Morrison 2007, p.139).

Despite the peremptory significance of the results obtained, *not all substantial conclusions of Margaret Morrison's book can be agreed* upon. I would like to commence my critical scrutiny of Morrison's results with an indication of recent discovery at the Swiss Large Hadron Collider of the celebrated 'Higgs boson'. It put an end to the chewing the cud of empirical adequacy of the electroweak theory, as well as the so-called 'Standard Model' of elementary particles, embracing, by the way, the strong interactions. The discovery was highly appreciated by the scientific community: a humble Scottish physicist Peter Higgs got his Nobel Prize in Physics (2013).

But a stronger thesis to be disputed consists in Morrison's assertion that the electroweak theory

"itself doesn't provide a full account of how the fields are mixed by furnishing direct values for the Weinberg angle (i.e. the degree of mixing is not determined by the theory). More importantly, *the theory doesn't determine in any strict sense that the fields must be mixed*" "Morrison 2007, p. 136; my italics).

In my view, the daring tenet that the theory in *no way* determines the fact that the fields should be mixed is too strong a statement to be accepted. It is grounded on a one-sided consideration of the encounter of the electromagnetic and weak theories in the theories of Fermi and Feynman and Gell-Mann. Forsooth, *empirically* ('phenomenologically') the interactions of electromagnetic and weak fields were not then fixed; another thing - the so-called 'encounter paradoxes' expressed in the appearance of divergences of the corresponding field quantities and the non-renormalizability of the Fermi and Feynman and Gell-Mann theories. This circumstance is well known and presented in the standard textbooks on electroweak theory, which were cited above (see, for instance, Greiner and Muller, 2009).

Nevertheless, in my judgment, no less convincing argument is the opinion of one of the celebrated researchers who received the Nobel Prize *exactly* for the unification of electromagnetic and weak interactions - professor of the Massachusetts Institute of Technology Sheldon Glashow. In his renowned Nobel lecture, '*Towards a Unified Theory - Threads in a Tapestry*', a graduate student of the ardent supporter of the unification — Princeton professor Julian Schwinger — shrewdly stressed that

"Of course, a gauge theory of weak interactions *alone* cannot be made renormalizable. For this, the weak and electromagnetic interactions *must be unified*. Schwinger, as early as 1956, believed that the weak and electromagnetic interactions *should* be combined into a gauge theory. The charged massive vector intermediary and the massless photon were to be gauge mesons. As his student, I accepted this faith. In my 1958 Harvard thesis, I wrote: "It is of little value to have a potentially renormalizable theory of beta processes without the possibility of renormalizable electrodynamics. We should care to suggest that a fully acceptable theory of these interactions may *only* be achieved if they are treated *together*" (Glashow 1981, p. 1321; my italics).

The Maxwellian electrodynamics, the GTR, the QED, and the electroweak theory were solidly put up via consequent synthesis of partial theoretical schemes. Moreover, in each of the synthetic theories, a 'neutral language' was worked out to display the encountered theories. In Maxwell's synthetic program, this is the blunt language of hydrodynamics and vortex theory (see Nugayev 2015 for details); in GTR- the punctilious language of Riemannian geometry (Nugayev 2017); in the electroweak program - the peremptory language of the Yang-Mills gauge theory. It was the 'sleight-of-hand' use of the 'neutral language' that made it possible to compare the material to be combined, turning it into partial theoretical schemes for subsequent generalization into a global theoretical scheme.

Other facets of the particular origin of the electroweak theory are unfolded when comparing this process with the genesis of Maxwellian electrodynamics and the origin of GTR. In particular, the basic ideal GTR model (or, in terms of Vyacheslav Stepin (2005), the fundamental theoretical scheme) was constructed by *synthesizing* less abstract theoretical models (or 'partial theoretical schemes') of Abraham, Nordström, and Einstein (see chapter I). In this case, both the partial mod-

210

els themselves and the basic model were crossbred (hybrid) models that combined the main features of all theories encountered, although to different degrees. Approximately the same epistemic scheme is also present in the case of the electroweak theory. The only substantial difference consists in that in the GTR case the lead of the heuristic principle that guides the construction of sequences of less and less hybrid models was played by the so-called 'principle of equivalence'. In the electroweak case, the role of this principle was famously fulfilled by the *'gauge principle'* - the longing to get as close as possible to the desired 'ideal' (Salam) - the Yang-Mills gauge theory.

The task of bringing to fruition the further stages of the development of the electroweak theory was practically reduced to creating its adequate gauge model by searching for the corresponding symmetry group. The next step consisted in tentative constructing the corresponding Lagrangian, amending it to achieve its invariance concerning gauge transformations, proving its renormalizability, comparison with experiments, and so on. In the general case, one can enunciate the 'reduction' of the electroweak theory to the Yang-Mills gauge theory, so that the actual method of synthesis of electromagnetic and weak interactions can be more aptly described as *'reductionist-synthetic'*.

The second component of the positive heuristic of the synthetic program for creating an electroweak theory was the permanent 'look back' at electrodynamics, the implementation of QED as a model (a 'paradigm') for conjuring a successful renormalizable gauge theory up. For this, not only and not so much the provisions of a successful theory were directly borrowed, but also the methods (for instance, the fine technique of Feynman diagrams).

Moreover, like the GTR, the electroweak theory was put up due to the sequential synthesis of systems of various hybrid objects, the first of which are intermediate vector bosons W^+, W^-, Z^0. More precisely,

"these three charges — together with the electric — form four components of a 'single' entity, each component transformable one into the other, through the operations of the group structure SU (2) x U (1) acting on an 'internal of symmetry space '" (Salam 1984a, p. 316).

It was these crossbred objects, constructed at the same time from diverse encountered theories, that carried out the transfer of properties from one mellow theory to another and the corresponding *modification* of these theories. Let us recall Planck's relentless dictum: *must be modified to remain compatible'*. It means, in particular, that not only the theory of weak interactions but also the QED should be modified in the encounter. However, namely, that was underscored by Abdus Salam,

"The most spectacular prediction, verified last year at the Stanford Linear Accelerator Centre was the following. If indeed the weak nuclear force is nothing but a different facet of a basic force whose other facet is electromagnetism, electromagnetism - the force between electrons and protons - when carefully examined *should show some characteristics which one had in the past associated with the weak nuclear force only.* One such characteristic is the distinction force experienced by left-spinning versus right-spinning electrons. The SLAC experiment, measuring this deflection with an accuracy never before attempted, demonstrated that left-spinning electrons indeed are deflected one part in ten thousand times more than right-spinning electrons when scattered off heavy water. To one part in ten thousand − just as the theory predicted − what was previously thought of as the distinct nuclear force intrudes into the hitherto separate domain of electromagnetism, clinching the hypothesis that the two forces are indeed facets of one basic fundamental force and are intertwined one with the other "(Salam 1984b, p. 304; my italics).

Nevertheless, there is a jolly significant difference between the reductionist-synthetic method of constructing the electroweak theory and the Maxwellian one. Maxwell's unification of electromagnetism and optics turned out to be incredibly exact, as a result of the introduction of a basic crossbred object — the displacement current — by calculating the speed of light from electric and magnetic constants (c $= 1/\sqrt{(\epsilon\mu)}$.

On the contrary, the Glashow – Salam- Weinberg bland combination of electromagnetic and weak interactions led to the introduction of the notorious Weinberg angle, which describes only the *mixing* of weak and electromagnetic fields. Alas, the Weinberg angle is not calculated in theory, but determined empirically, from the experiment.

5.4 Part three. Mixed Conclusions

First, this hallmark demonstrates the *incompleteness* of the electroweak theory. As Lev B. Okun shrewdly noted, the fact that weak and electromagnetic interactions are characterized by the same charge e is the visual expression of the fact that the standard model of electroweak interactions constitutes a unified theory of weak and electromagnetic interactions, forsooth. But

"The fact that the model has a free parameter θ_μ, which is theoretically not fixed and not predicted, means that a unified theory *has not yet been accomplished*. (From experience ... $\theta_\mu \approx 30^0$). The presence of the free parameter θ_μ is a consequence of the fact that the symmetry group of weak interactions is a direct product of two simple groups: SU (2) and U (1). We could get rid of this arbitrariness if we demanded that both of these groups be subgroups of some larger group" (Okun 1990, p.204; my italics).

The verdict of the Russian academician well complies with the opinion of the American Nobel laureate, famously professed in the renowned Nobel lecture '*On the way to a unifying theory - threads in a tapestry*':

"Let me stress that I do not believe that the standard theory will long survive as a correct and complete picture of physics. All interactions may be gauge interactions, but surely they must lie within the unifying group" (Glashow 1981, p. 1420).

It is still possible that the accomplishment of the ultimate synthesis of electromagnetic and weak interactions will nevertheless be given within the inextricable framework of such actively developing areas as supersymmetry, supergravity, and superstring theory (see, for instance, Maldacena, Strominger, Witten, 1997).

THE EPILOGUE. AN ENTICING RECONCILIATION OF GENERAL RELATIVITY AND QUANTUM FIELD THEORY

The GTR contrivance fatally led to the problem of the reconciliation of GTR and QFT. This problem is not solved yet, and Stephen Hawking famously called it 'the most urgent problem facing theoretical physics today'.

Although the first efforts to reconcile GTR and QFT started from the very moment of GTR creation (theories of Weyl, Cartan, Kaluza & Klein), the substantive progress was attained quite recently. At least at the first stages, it surprisingly seems to have nothing to do with superstrings, supergravity, and so on. Following my epistemic model, I reckon that the real progress in QFT and GTR unification commenced with the hybrid objects' construction in a brand new crossbred realm called the '*Quantum Field Theory in Curved Spaces*'. (Birrell & Davies had issued in 1981 the first review under the same title). The crossbred realm embraces at least two important domains – (BP) black hole physics and (IC) inflationary cosmology. Let me examine these modern examples and their alluring interrelation in more detail.

(**BP**) The most apparent example of genuine cross-theory is Hawking's path-breaking 1975 paper on the particle creation by black holes. Yet the real story started earlier, with Jacob Beckenstein's pioneering inquiry into the black hole entropy. His paper turned out a masterpiece artfully combining deep physical intuition with outrageous flies of imagination so rare in the dark waters of mathematical speculations.

Beckenstein's salutary quest had commenced with a striking analogy between entropy and the surface of the black hole horizon. From the fecund information theory standpoint, entropy and horizon of a black hole are the quantities of the same kind. If a particle gets under the horizon of a black hole, the surface of the hole increases due to the mass addition. At the same time, entropy increases also, since one cannot look under the horizon to know the state of the particle there. Consequently, the area of the horizon should be equated to entropy with some proportionality coefficient.

J.M. Bardeen, B. Carter, and S.W. Hawking (1973) generalized and clarified Beckenstein's startling, yet semi – intuitive, ideas. They envisaged the profound analogy between black hole physics and thermodynamics in more strict and mathematically sophisticated language. The fine expressions were sedulously derived for the mass of a stationary axisymmetric solution of the Einstein equations containing a black hole surrounded by matter and for the difference in mass between two neighboring solutions.

"Two of the quantities which appear in these expressions, namely the area A of the event horizon and the 'surface gravity' k of the black hole, have a *close analogy* with entropy and temperature respectively. This *analogy suggests* the formulation of four laws of black hole mechanics which correspond to and in some ways transcend the four laws of thermodynamics" (Bardeen, Carter, Hawking 1973,p.161; my italics).

Now GTR and thermodynamics were merged into a general scheme, and the corresponding conspicuous results showed up rather quickly. According to the Second Generalized Law of Thermodynamics, formulated by Beckenstein and Hawking, the flow of entropy across the event horizon of a black hole should increase the area of the horizon A. The important quantity "entropy + some multiple of A" never decreases. Yet if one equates the surface gravity of a black hole k to temperature T, a startling paradox occurs. If a black hole is immersed in the sea of blackbody radiation at a *lower* temperature (than that of the black hole), the Second Law will be necessarily violated! It should be stressed that the only way out to avoid the paradox is to admit that a black hole should emit particles at a steady rate with a temperature equal to $k/2\pi M$. Namely, that committed Stephen Hawking (1975) by shrewdly considering quantum-field effects. At least several lessons can be learned from the story. Firstly, three mellow theories were involved in the occurrence of the cross-contradiction: thermodynamics, GTR, and QFT. Besides, QFT is laid out as a necessary link connecting GTR with thermodynamics. Without QFT, GTR cannot be properly reconciled with thermodynamics and statistical mechanics. The conspicuous story looks like that: classical thermodynamics 'knows' something about the quanta, as if the quantum theory in some curious way is 'contained' in the thermodynamics. In the light of the already told (see also chapter II for details), this conclusion does not seem jolly strange; the quantum theory already is a wayward fusion of thermodynamics, mechanics, and electrodynamics.

"This has perhaps contributed to the idea that gravity has a somehow *deeper* relationship to thermodynamics than to other branches of physics" (Gould 1987, p. 449; my italics).

From the very beginning, the QFT in Curved Spaces met with many paradoxes caused by the lack of proper agreement between GTR and QFT (Fulling 1973; De Witt 1975). The stout notion of a 'particle' unexpectedly turned out to be observer-dependent. Moreover, exasperating conformal anomalies suddenly occurred; various renormalization methods furnished diverse results for vacuum energies; the queer negative energy fluxes began to fill the space with GTR, and many other such bizarre contraptions began to bother poor theorists and continue to puzzle them. It is important for the developed epistemic model that at least one of the leading black-hole boffins had professed: the cause of all these paradoxes consists in the semiclassical character of black-hole evaporation theory.

"The gravitational field is treated 'classically' (as a space-time metric) there, while its sources are described in terms of quantum field theory. The basic equations of GTR, $G_{\mu\nu} = T_{\mu\nu}$, contain *incompatible* quantities on the two sides of the equals sign. Furthermore, this incompatibility has consequences even at the present experimental scale" (Unruh 1980, p.154; my italics).

One of the most fascinating crossbreeds here is the whimsical QFT vacuum When it was intruded to the GTR derivative system (the black hole physics), it did acquire a new and rather queer property – 'to produce infinitely great gravitational field'. However, the common methods of its regularization (the so-called 'cut-off procedures') were elaborated for flat spacetimes only. Whereupon they are useless in strong gravitational fields.

The introduction of the QFT vacuum restructured the relations between the GTR objects significantly, and the process of the global theory construction inevitably began. Besides the black holes, various inflation scenarios are the indispensable stages of the global theory construction too. The enticing creation of the synthetic global theory seems to follow the promising lines of Einstein and Bohr.

(**IC**) The whimsical QFT vacuum gained particular urgency with the advent of quantum theory when it was realized that all fields of Nature are endowed with irreducible 'zero-point energies', even in their vacuum state. It is quite important that the notion of zero-point energy is a pure result of quantum theory and has no classical counterpart. Incidentally, it was introduced by Planck in an address to the German Physical Society in February (Planck 1911).

Grounded on the ideas underlying his second theory (Kuhn 1978), Planck had punctiliously calculated the average energy of an oscillator vibrating with frequency ν and with absolute temperature T:

$E = h\nu/1 + h\nu/\exp(h\nu/kT) -1.$

The value of the quantized energy levels of an oscillator could be written as

$E = (1/2)(E_n - E_{n-1}) = (n+1/2) h\nu,$

where n= 0,1,2,... At T= 0 the average energy is not zero but equals the finite energy $(1/2) h\nu$.

For a long time, Planck's bold hypothesis was considered dubious and controversial. The point of view that zero-point vacuum energies and fluctuations are indeed real was first evidenced by the so-called Casimir effect going back to 1948. Casimir predicted from the QFT basis that there exists an attractive force between two perfectly conducting parallel plates $F \approx d^{-4}$, where d is the distance between the plates. According to Casimir, the difference between the zero-point energies in a pure vacuum and the presence of the plates would be measurable and it was (Sparnaay 1958).

Initially, the Casimir effect was not considered relevant for cosmology. For a long time the cosmological constant λ, famously introduced in 1917 by Einstein to keep the Universe stable, and the quantum mechanics of the vacuum lived separate lives. The situation only changed in the late 1960s. Inspired by a revival of interest in cosmological models with a positive cosmological constant, Yakov B. Zel'dovich (1968) stressed the significance of λ in the context of QFT. He also elicited the first version of what would be known as a cosmological-constant problem.

"With Zel'dovich's work two historical strands were finally joined; the quantum vacuum and the energy density related to the cosmological constant" (Kragh and Overduin 2014, p.57).

Zel'dovich's influential works initiated the contrivance of another important crossbred realm – the inflationary cosmology. Inflation is the idea that the very early Universe may have been dominated by vacuum energy, giving rise to a short period of vastly accelerated expansion. However, inflation did not obtain wide recognition until high-energy particle physicists became involved. Stimulated by the advances of electroweak unification, they started to examine the implications of spontaneous symmetry breaking for cosmology, with the attendant possibility of a false vacuum state. Also, up to now the most compelling argument for inflation consists in its potential to link the observed large-scale structure in the present-day Universe to antediluvian quantum fluctuations in the earliest moments after the big bang.

Yet ultimately, despite an apparent success of inflation models in addressing many conundrums of the early Universe, theoretical cosmologists were reluctant to entertain the idea that vacuum energy might also dominate the Universe later until they were obliged to do so by astronomers. Improvements in detector technology, image analysis, and other factors led to the definitive detection of dark energy in 1998 using the magnitude-redshift relation for Ia supernovae. The quick convergence of two independent teams to a similar conclusion played a large part in convincing the cosmological community of the reality of a vacuum-dominated accelerating Universe. Yet we do not know why λ is so much smaller than calculations based on QFT would suggest

(Perlmutter 1999).

Thus, vacuum plays a pivotal role in the entire modern crossbred realm, being a steady global object of the future theory: 'making everything out of nothing' (Robert Weingard) during the process of spontaneous symmetry breaking. Incidentally, the Quantum Vacuum played the leads at each stage of the unification of the four fundamental forces of nature - in Higgs mechanism establishment (Higgs vacuum), in fifth coordinate compactification in Kaluza-Klein models (the Casimir effect), in the Green & Schwartz superstring model of Quantum Gravity free of gravitational anomalies.

Let us turn, at last, to the most thrilling and fresh example – modern super-unification theories, embracing the Superstring Theory. In what sense this theory reconciles the GTR with QFT, becoming the celebrated 'Theory of Everything'?

To begin with, the scientific research program of adequate superstring theory construction can be described using internalist Lakatosian standards embracing positive and negative heuristic, protecting belt of auxiliary hypotheses, etc.

In particular, its hard core contains the following propositions.

1. The carriers of the fundamental interactions are nonlocal.

2. The interrelations between bosons and fermions are described by the Supersymmetry Principle.

3. Space-time is multidimensional in Theodore Kaluza's whimsical sense.

As usual, the elements of the hard core were borrowed from other branches of elementary particle physics. In particular, initially, the principle of nonlocality had been worked out in hadronic physics. Gluonic fields appeared to be concentrated along the lines that connected the famous 'quarks'. Eventually, it leads to the models of mesons consisting of one-dimensional strings connecting a quark-anti-quark pair. Despite some obvious successes, the string theory advancement met with considerable impediments. According to the most dangerous one, the spectrum of the particles predicted by such theories should contain the notorious *tachyons* (that propagate faster than light).On realizing that, Scherk & Schwartz made a small revolution proposing a distinctive hypothesis that the space-time dimension is equal to 10.

It had changed the program's hard core substantially. Furthermore, the heuristic was altered by the cool assertion that all the unification should be effected in full analogy with that of Theodore Kaluza by putting all the new fundamental interactions into new dimensions. As always, the trick had eliminated the old difficulties forsooth but had created the new ones. All appearance, the transition to superstrings was not ad hoc$_3$ and did not spoil the spirit of the research program significantly. However, it did not lead to new predictions that were experimentally confirmed. The purpose of the following part is to strengthen arguments in support of the tenet that the genesis and character of the newly discovered difficulties cannot allow one to allege that the global theory unifying all the four basic interactions have been already created.

The reverie to unify gravity and electromagnetism is as old as mountains. However, in 1921, the German mathematician Theodor Kaluza proposed a dainty unified theory of gravitation and electromagnetism by assuming that space-time is 5-dimensional with one timelike dimension and four spacelike ones. The main merit of the theory consisted in the possibility of treating the newly born geometrical quantities as the electromagnetic - field potentials. Moreover, these were achieved in the following wayward way.

The line element is commonly $ds^2 = g_{ij}dx^idx^j$, where for 5-dimensional space-time i, j = 1, 2,3,4,5. The metric tensor g_{ij} is a matrix consisting of a usual 4-dimensional metric tensor $g_{\alpha\beta}$ (α,β =1,2,3,4), two vector fields A_α and A^β, and jolly unpleasant - dark and obscure - component $G_{55.}$ After putting it equal to -1 and postulating a whimsical ' *cylinder condition*', according to which the usual 4-dimensional metric tensor should have a vanishing derivative concerning the newly introduced fifth dimension, both vector fields A turned out very much alike the 4-potentials of the electromagnetic field. At the same time, 5-dimensional Christoffel symbols that contain the first derivatives concerning all the 5 coordinates transpired to be consisting of 4-Christoffel symbols, which are analogous to gravitational field stresses, and of the Christoffel symbols that contain the index 5. The latter reminds one of the electromagnetic field stresses. If one identifies them with the electromagnetic stress-tensor $F_{\alpha\beta}$, he/she can derive some thrilling consequences that were called by the enthusiasts the 'Kaluza-Klein miracles'. Let us dwell on them.

1. The first miracle can be produced, if one equates the 5-velocity of a particle dx^5/ds to e/m, where e is the particle's charge and m it's mass. Then 5-dimensional geodesic equations turned out to consist of 4+1 equations. The first four of them are the usual equations of motion of a charged particle in the gravitational field in GTR.

2. *The Pivotal Miracle.* The 5-dimensional Einstein field equations $R_{ij} - (R/2)g_{ij} = \kappa T_{ij}$ (where T_{ij} is now exclusive of electromagnetism) unfathomably reduce to the usual Einstein equations in 4 dimensions and the common Maxwell equations, and, alas, to the exasperating equation on unpleasant G_{55}. (That should not deserve the attention of bold inventor of unified field theory).There are two more miracles - the appearance of the electromagnetic stress-tensor on the right side of the equations and the explanation of the electromagnetic field gauge invariance. However, they appear too modest in comparison to the Pivotal Miracle that attracted even Einstein for several decades of his life.

Kaluza's original inspiration was rendered a firmer basis by Oscar Klein in 1926 - mainly in geometrical respects - and finally, the odd contraption was called the *Kaluza-Klein theory*. Besides Einstein, the theory was further advanced by Viktor Fock, Jakov Frenkel, and Louis de Broglie. Hence, the attraction of the theory had many ups and downs until the 80-th, when Scherk & Schwartz managed to contrive a bestseller out of the old physical anecdote. However many (mostly heroic) efforts devoted to elaborating the theory had elicited some deficiencies pointing out that the hypotheses of Klein and Kaluza made the unification program too ad hoc. Viz.,

1. Why one adds to 4 only one dimension? And not two? Or more?

2. The 'cylinder condition' is apparently of very artificial origin: $\partial g_{\alpha\beta}/\partial x^5 = 0$. Whereupon why should the 4-dimensional metric

tensor with components describing the gravitational-field stresses in all the points of space-time obey such a tough condition?

3. Why it is the fifth dimension that is not observable? And not the third one? Klein's earlier suggestion that space-time is periodic in the new fifth dimension and that 5-spacetime is to be thought of as homeomorphic to a tube, the direct product of 4-spacetime by a circle with radius x^5 satisfying $0 < x^5 < 2\pi r_5$, is not the true explanation. It is only a statement of the fact that the fifth dimension is closed. In the Kaluza-Klein dizzying approach, one simply asserts that the radius r_5 of the circle is jolly small - only a few orders of magnitude larger than the Planck length $hG/c \cong 1.6 \times 10^{-33}$ cm. Hence, the fifth dimension is not observed in everyday experience due to the confounded ad hoc hypothesis. Einstein's dictum – 'one must explain why the continuum is restricted to four dimensions' - is not satisfied in the Kaluza-Klein odd theory.

4. How should the G_{55} component of the metric tensor be comprehended? In the most general case, this component represents the scalar field analogous to Coulomb one. The Brans & Dicke efforts to insert in a consequent way this field into the GTR were not fruitful. They did not provide any new result save the assertion that the field's coupling constant is too negligible to be measured by modern experimental devices. Moreover, interpretation of this whimsical component *a la* Brans & Dicke is a new ad hoc hypothesis in turn. Yet if one (as Kaluza first did) asserts that $G_{55} = -1$, this will result in wild physical consequences (see Vladimirov 1989 for details).

Critical arguments (1) - (4) can be summarized as follows. Kaluza-Klein 'unification' of gravitation and electromagnetism grounded on the assumption that space-time is 5-dimensional and identification of the newly obtained geometrical quantities with electromagnetic-field potentials is too formal to be taken seriously. To insist that gravitational and electromagnetic fields are different parts of the same metric tensor, so that 5-dimensional Einstein equations can be reduced to 4-

dimensional Einstein equations and 4-dimensional Maxwell equations, is in fact to maintain that gravitational and electromagnetic fields are of tensor origin. However, the latter proposition is a platitude. Einstein and Maxwell's equations were obtained *before* the Kaluza-Klein deft theory. Yet this theory tells *nothing new* neither to experimentalists nor to theoreticians. Hence, in my view, the Kaluza-Klein theory represents ad hoc_1 and ad hoc_2 hypothesis in Lakatosian terms in the scientific research program of unified field theory construction.

Once upon a time in America Richard Feynman sarcastically noted that it is quite easy to write out the most fundamental equation that represents the most general physical law that per saltum 'unifies' all the known interactions. The law is $\Box F = 0$, where the operator $\Box$ and function F vary for any novel field under consideration. For instance, for classical mechanics, $\Box = 1$, and $F = f - ma$, for classical electrodynamics $\Box$ is the D'Alembert operator and F are the corresponding field quantities,etc.

Why is this dubious unification formal? – Because any real - and not formal - unification should result in the construction of the Fundamental Theoretical Scheme or basic ideal model of the process under consideration (see Stepin 2005 for details). Having had rationally reconstructed (together with Lev Tomilchik) the process of electricity, magnetism, and theory of light fusion, Stepin did notice some peculiarities of the unification process, that have general significance, in my view, for any process of successful scientific synthesis. They can be summarized in the following way.

1. Any mellow scientific theory is a set of propositions describing the relations between different *sets* of abstract theoretical objects. The abstract objects of a mellow theory belong either to a subset of basic objects or to a subset of derivative ones. The latter is constructed from the fundamental objects by some rules belonging to a certain paradigm (in Kuhn's sense).

2. To provide an ability to predict new phenomena, the objects of the global system should be *operationally* defined through the ideal measurement procedures. The connection to lower-level abstract objects should be specially disclosed.

3. A mellow theory is so structured that the throng of its multifarious models is disseminated over at least three following interconnected *levels* (Stepin 2005; see also the corresponding comments in Vihalemm and Müürsepp 2007).

   (a) The *top* level of the *basic* ideal model where 'the Fundamental Theoretical Scheme' resides.

   (b) The *intermediate* level of the *subordinated* ideal models (or 'the Partial Theoretical Schemes') constructed out of the basic one according to certain (often-tacit) rules.

   (c) The *inferior* level of the *'Empirical Schemes'* that can be approached through the level of partial theoretical schemes, on the one hand, and via the wayward and often disputable 'generalization' of the controversial empirical data, - on the other.

The links between all the three levels of an established mellow theory should be sufficiently solid and stout ones. Their rigidity allows one to connect a prediction referring to the upper level with all the underlying levels of a mellow theory. Hence, it allows one to construct an experimental device capable to check the prediction. A new result, obtained in the course of the advancement of mathematical apparatus, influences all the levels of a mellow theory immediately. Hence, a mellow theory can predict, and the predictions can be successfully verified. A mellow theory obtains the status of an established one when at least some of its predictions are posited to be successful. It exhibits that the system of basic objects is complete, and all the necessary links between all the three levels are made sufficiently robust. Owing to the subtle structure of a mellow theory, the global theory contrivance appears to be a slow, adamant, and consequent ascent from the lower levels up to the top ones. Any transition from the lower level to the upper one is impossible until the construction of all the lower-level models is accomplished.

A violation of (i) - (iii) should violate the *rigidity* of the levels' connection and consequently should lead to a decrease of the unifying theory's predictive power.

The proponents of Kaluza-Klein theory can retort that (i) - (iii) belong to the classical stage of physics development. They can fend that modern physics began with Einstein's reconciliation of electrodynamics, mechanics, and thermodynamics in 1905 (chapter III) and his 1915 startling unification of Special Relativity and Newtonian Theory of Gravity. Modernity in physics had culminated in the creation of Quantum Electrodynamics. This stage is characterized by the domination of the mathematical hypothesis method, when one tries to guess the basic equations or the basic unifying laws with a help of such regulative principles as Beauty, Simplicity, etc. Only after establishing the basic laws, one has to provide their empirical justification and to link the system of basic objects to those of the lower level. The most vivid example is the creation of Quantum Electrodynamics when special Bohr-Rosenfeld ideal-measurement operations were elaborated.

The proponents of Kaluza-Klein's theory can point at GTR as a paradigm of theory construction common for the XX century. However, the counter-argument can be put forward according to which the GTR contrivance had merely a distant resemblance to the mathematical hypothesis method. This theory was constructed by the generalization of partial crossbred theoretical schemes (see the example of chapter I and Nugayev 2018). Fortunately, the GTR consists not only of Einstein's equations. It includes the so-called 'weak-field approximation in GTR', STR, and Newton's theory of gravitation as well.

However, let us return to the superstrings. As is well known, supersymmetry is a symmetry between bosons and fermions. Even in the simplest supersymmetric theory, the usual 4-dimensional space-time is enlarged to form the superspace, any point of which has 8 coordinates: 4 usual x, y, z, t, and four new ones belonging to the so-called Grassman's algebra. In the first approximation, the usual coordinates correspond to bosons, and the Grassman ones correspond to fermions. If the usual space-time admits a 10-parameter Poincaré group, the superspace admits 14-parameter enlarged Poincaré group in which to usual transformations the supertranslations are aptly added. In supersymmetric theory, all the fields are exchanged on superfields that depend, in the case of the simplest supersymmetry, upon 8 variables, and in the case of N-enlarged supersymmetry - upon the 4+4N ones.

Supersymmetric Lagrangian, as always, is defined in the form containing invariant square expressions composed from superfield's first derivatives along with all the 8 coordinates. Superaction is defined as an integral of the Lagrangian along with all the variables.

All these are related to *global supersymmetry*. The next step in the theory advancement should consist in *localizing* the supersymmetry, i.e. in introducing the dependence of supertranslational parameters upon the usual 4-coordinates. Just as the localization of usual Poincaré group parameters leads to GTR creation, localization of 14-parameter enlarged Poincaré supergroup results in the contrivance of supergravity. The number of additional variables N varies in these theories from 1 to 8. Maximally enlarged N=8 theory of supergravity consists of one spin -1 field, eight fields with spin $3/2$, 28 spin 1 fields, and 70 spin -0 fields. It is no surprise that supergravity proponents did not decide yet what to do with all these night treasures and with what fields should the supermultiplets be identified. Besides, who can provide the superselection rules then?

Nevertheless, "the process had begun already" and supergravitational Kaluza-Klein theories with n usual and m Grassman coordinates were proposed. Especially interesting the supergravity theories appeared in 11 dimensions since 11 is the minimum number necessary for introducing the gauge Great Unification Group SU (5) x SU(2) x SU(1).

Eleven is the least space-time dimension to include electromagnetic, strong, and weak interactions. Moreover, namely, 11 dimensions admit compactification of additional 7 dimensions. However, how,in what ways? Let us inquire into the methods of Kaluza-Klein supergravity theories construction.

These theories are created by a simple generalization of 5-dimensional Kaluza-Klein theories on the case of N = 4+D dimensions. However, the method is based on the assumption that the Einstein equations are valid for D dimensions. The ground state from the very beginning is chosen as $Ì^4$ x B^D, where B^D admits a group of isometries generated by D Killing vector fields, and not as $Ì^{4+D}$. Yet the latter expression should be valid to make the unification correct.

Matching expressions for metric are derived. In complete analogy with 5-dimensional $\grave{I}^4 \times S^1$, a group of isometries B^D should show it-self up as a group of gauge symmetries of the fields existing 'within' $\grave{I}^4$. Indeed, the Kaluza-Klein genuine approach was exhibited in the first part of this part. The gauge invariance of the electromagnetic field turns out to be a consequence of the fifth coordinate special role. Gauge invariance appears to be an expression of 5-dimensional sym-metry in a 4-dimensional world. Space-like Killing vector field $\partial/\partial x^5$ generates isometries that appear as the U (1) gauge symmetry in $\grave{I}^4$.

Hence, in the most general case, one can always choose $\hat{A}^D$ to unify gravity with *any* gauge group. They say that $\hat{A}^D$ admits the isometries group G generated by D Killing vectors. In 4-dimensional space-time, G will look like a non-abelian gauge group. Gauge invariance is a sim-ple space-time invariance in multi-dimensional space-time. In complete analogy with the Kaluza-Klein theory, the metric tensor in 4+D di-mensions can be written as 4-dimensional metric plus 4D components of gauge field quantities plus a pile of scalar fields. $\grave{I}^4 \times \hat{A}^D$ will be a su-perspace now. The opportunity of such a representation of the metric in 4+D dimensions is due to the generalized 'cylindrical conditions'. They appear now as specific restrictions shaping group generators, i.e. the Killing vectors. They let the generalized Lagrangian to be written as a sum consisting of two parts (Chyba 1985).The first part is a usual Lagrangian of a free field in 4 dimensions and the second one is the Yang-Mills one describing the gauge fields under consider-ation. From these Lagrangians by standard quantum-field methods the 4-dimensional Einstein equations, as well as the Yang-Mills ones (including the equations of Maxwell, Klein-Gordon-Fock, and Dirac et al.) are derived.

Since the procedure described is a simple-minded generalization of a 5-dimensional case, it is no wonder that *all* the Kaluza-Klein celebrated miracles relentlessly remain.

Alas, all the drawbacks remain too. Also, the new ones appear: one has to invent the paraphernalia to compactify now three dimensions, and not one. And of course, the question occurs: why the initial dimension of space-time is equal to 11? And why 7 of them should compactify later? Why all of the compactified dimensions are spatial ones?

Incidentally, I conclude that the dimensionality problem is not solved by an anthropic argument. It is masterfully reformulated to compose an illusion of the answer. Moreover. We have additional scalar fields connected with 4+D metric components. What shall we do with them now? Where can one find so many Branses and Dickes to look for their places in the Universe? Also, some novel problems occur that were absent in the genuine Kaluza-Klein theory. The first one is the problem of vacuum stability and the second one consists in that the $\grave{I}^4 \times \hat{A}^D$ manifold is not a solution of the Einstein equations in vacuum. A significant impediment spoils the picture significantly.

One can conclude that the main Kaluza-Klein shortcoming - the lack of new content in comparison with the older theories - remains in the generalized multi-dimensional versions of the original 5-dimensional theory. The unified case contains the same information as the number of theories before unification. Nothing new. The origin of the obstacle consists in the following. Multidimensional generalization of the Kaluza-Klein theory is as *trivial* as the 5-dimensional original version since it represents a simple translation of the gauge invariance idea into the geometrical language. It contains nothing new in comparison to the Yang-Mills fine equations, just as the translation of Conan Doyle's well-known novel from English into Russian cannot transmute the honest detective Sherlock Holmes into the criminal professor Moriarty and vice versa.

To summarize, the basic argument for superstring theory consists in that it has *no* solid alternatives. For instance, the loop-gravity approach reduces to innumerous discussions of the multifarious opportunities of unification of 4 basic interactions without proposing a reliable unification strategy. On the contrary, within the superstring theory, certain steps are taken in the process of erection of the Global Theoretical Scheme, which necessarily contains an actual global theoretical object – an oscillating superstring. All the basic theoretical objects of GTR and QFT ought to be deftly constructed from it. Secondly, one of the most convincing arguments in favor of a theory is its ability to predict the facts that were not used in the process of its construction. Likewise, a theory can be supported in a theoretical way. For instance, it can reveal important novel theoretical interconnections between the different parts of the same theory or new substantial links between the different theories as well. Thus, the theoretical support may exceed even the sheer empirical corroboration. In that way, the superstring theory rendered the whole set of theoretical achievements the most important of which consists in the following 'fact'. It reproduced theoretically and even managed to explain the Beckenstein - Hawking equation relating the entropy of a black hole to the surface of its horizon. Viz.,

"The explanation of black hole entropy in terms of microscopic states is widely regarded as one of the benchmarks for theories of quantum gravity. The analogy between the laws of black hole mechanics and the laws of thermodynamics, combined with Hawking's effect, suggest assigning to a black hole of area A the 'macroscopic' (or 'thermodynamics') entropy $S_{macro} = A/4$ (1.1) S_{macro} depends on a small number of parameters which can be measured far away from the black hole and determine its 'macroscopic' state: the mass M, the angular momentum J and its charge Q concerning long-range gauge forces...

A theory of quantum gravity should be able to specify and count the microstates of the black hole, which give rise to the same microstate. If N states are corresponding to a black hole with parameters M, J, Q, then the associated 'microscopic' or 'statistical' entropy is $S_{micro} = \log N$ (1.2)

By the analogy to the relation between thermodynamics and statistical mechanics, it is expected that the macroscopic and microscopic entropies agree" (Mohaupt 2005, p.3).

Nevertheless, despite more than 30 years of stubborn research, the superstring theory remains not only *empirically unconfirmed* but also *theoretically unfinished*. We are still unaware of the regions of the Universe that can be explained by strings. One cannot even speak of certain predictions of the behavior of any cosmological objects.

The long list of the numerous difficulties of the superstring theory can easily be extended (see for instance Smolin 2006; Woigt 2006; Hossenfelder 2020) but it is already evident that the very existence of consequent, verified Theory of Everything is out of the question. However, one can ascertain the continuation of the reconciliation of GTR and QFT commenced by Beckenstein and Hawking. In this regard, the main achievement of the superstring theory consists in that it changed GTR to reconcile it with QFT (Michael Green &John Schwartz 1984). Moreover; the string theory furnishes the mechanism demonstrating how all the four fundamental interactions transpire from the same beginning – an oscillating superstring. According to Edward Witten,

"Respectively, one can expect that the basic notions of the quantum theory should be changed to provide the full integration of gravitational physics into the quantum theory" (Dawid Richard 2013, p.28).

Why not? And how many (super)strings can fit on the tip of the needle?

SUBSTANTIAL CONCLUSIONS.

Briefly, the main results of the monograph can be summarized as follows.

1. A guileless, yet lucid and comprehensible **epistemic model** aimed at an analysis of the sober reasons for mellow theory change in science is erected. According to the model, *an origin* of a scientific revolution consists not in well-trodden discrepancies between bold theoretical speculations and stubborn 'hard facts', but in deadly contractions of 'old' mellow research traditions with each other. '**In this mutual adjustment lies the germ of their further development into a complete unity**' (Planck); the profound contradictions may be eliminated most effectively and radically in a more general 'global' theory.

2. It is exhibited that the dynamic creation of the global theory is being incessantly governed by strong internal tensions between the 'old' entrenched research traditions. Viz., the encounter of the traditions, their deep interpenetration, and subtle intertwinement entail the construction of a vast **hybrid realm** at first with a haphazard throng of crossbred theoretical models. Gradually, on consecutive soothing and eliminating the contradictions between the crossbred models, the hybrid system is sedulously arranged.

3. Up to a point, the epistemic model can be taken as rendering a subtle *mechanism* of a simultaneous profound transformation of the 'old' paradigms. The hallmark of the mechanism is the construction of the **crossbred systems** from the basic theoretical objects of 'old' mellow theories. These crossbred systems furnish the channels through which the accommodation of the 'old' traditions encountered

unstoppably commences. The accommodation ultimately leads to the installation of a medley of *crossbred* theoretical schemes. The medley will be ultimately processed and subsequently generalized to mold a conspicuous *structure* of a novel global theory. A mellow physical theory may be empirically successful if it represents a kind of a 'draft' that can be explicated by the results of new experiments that increase its empirical content. The global theory should become such a vehicle for the production of new propositions. For this purpose, it should reconcile and intertwine the 'old' classical traditions in such a way that they cease to contradict each other, so that newly- constructed fusion can successfully explain and predict new experimental data.

4. The philosophical background of the epistemic model is punctiliously unfolded embracing in the neo-Kantian touch Hilary Putnam's subtle **'internal realism'** and wayward **'coherence theory of truth'**.

5. An origin of the so-called 'old quantum theory' represented by Planck's startling 1897-1906 papers is revealed. It is manifested that the quantum-theoretical schemes were punctiliously constructed as 'crossbreed' ones composed of ideal models and laws of Maxwellian electrodynamics, Newtonian mechanics, statistical mechanics, and classical thermodynamics. In this pioneering medley, Ludwig Boltzmann's seminal ideas and subtle technique turned out to be crucial. Forsooth, deriving his blackbody radiation law, Max Planck had to take the experimental evidence into account. Yet it spurred him not 'to deduce from phenomena' punctiliously but to involve more theory instead. The blackbody solid experiments incited Planck to apply the statistical technique for proper arranging the radiation data in increasing portions. It is contended that Planck's startling discovery was by no means accidental. Planck's hybrid and intricate theories epitomized the commencement of an ambitious program of Maxwellian electrodynamics and statistical mechanics effective **reconciliation**.

6. It is contended that Einstein's 'annus mirabilis' constituted an acme of the second scientific revolution. To fathom in what felicitous ways Einstein's 1905 writings hang together one is bound to pay special tribute to his longed strive for unity evinced in incessant attempts to coordinate the profound research traditions of classical physics.

Though Einstein's craving efforts sprung out of Max Planck's pioneering attempts to comprehend electromagnetic phenomena through the lenses of conceptual structures of thermodynamics and statistical mechanics. It was Planck, who realized the cross-contradiction between 'the physics of material bodies and the physics of the ether' and outlined the first sketch of its withdrawal: the paradigms 'must be modified to remain compatible'. And it was Planck who took the first step in modifying the physics of the ether and contending that 'not only matter itself but also the effects radiated from matter' possess discontinuous properties, which can be characterized by a new natural constant: the elementary quantum of action. Einstein's part consisted in that he took the next step in modifying the second component of the encounter – the physics of material bodies. 'The modification in mechanics due to the principle of relativity is the introduction of a new universal constant, entirely foreign to classical mechanics, namely the velocity of light in an absolute vacuum' (Planck). **Einstein's foolhardy light quanta hypothesis and distinctive special theory of relativity turn out to be mere milestones of the unwinding of Maxwellian electrodynamics and statistical thermodynamics reconcilement research program**. The notorious conception of luminiferous ether was a substantial obstacle for Einstein's wayward statistical thermodynamics in which the pivotal lead was played by flagrant light quanta paper. Herewith Einstein was aware that his enticing light quanta hypothesis was too audacious to be taken literally. Hence he laid out his version of 'electrodynamics of moving bodies' in a markedly Machian/Duhemian, phenomenological way, and the basic relativity postulate was disposed of as originating from unfathomable and subconscious 'instinctive knowledge'. Forsooth, in his relentless criticism of the moldy research traditions of classical physics David Hume and Ernst Mach. influenced Einstein. However, when related to *creative* momenta, his 1905 reconcilement *modus operandi* was initiated by Mach's pre-eminent 'principle of economy of thought'. Yet the principle was taken in the subtle context of Stevin-Mach 'instinctive knowledge' wayward doctrine and with promising inspirations of the regulative spirit of Kantian and Duhemian non-representationalist epistemologies. It is manifested that for Einstein Duhem's confounded conventionalism was an indispensable bridge connecting and accommodating two quite disparate subjects of Machian crude empiricism and Kantian academic apriorism and transcendentalism. It constituted an indispensable means of saving special relativity postulates from scarce criticism for their inseverable connection with the foolhardy light quanta ideas.

The aim of the book is to take a step towards a complete description of Special Theory of Relativity genesis and acceptance, bringing some additional light on the intertheoretic relations between STR and other physical theories of the day. It should be stressed that the approach proposed enables to explain *why* classical mechanics and classical electrodynamics were "refuted" almost simultaneously or, in more suitable terms, why did the quantum and the relativistic revolutions both took place at the beginning of the XX-th century. It is exhibited that **the Quantum and the Relativistic revolutions were simultaneous since they had the common origin - the clash between the fundamental theories of the second half of the XIX-th century that constituted the 'body' of Classical Physics.** The revolution's most dramatic turning point was Einstein's 1905 light quantum paper, that laid the foundations of the Old Quantum Theory and influenced the fate of the STR too.

7. The mellow epistemic theory-change model, tried on the old quantum theory and special relativity genesis and advancement, is unfolded to apprehend general theory of relativity (GTR) genesis. It is demonstrated that the dynamics of GTR construction was largely governed by internal tensions of such research traditions as special relativity and Newton's theory of gravitation. The research traditions' encounter and interpenetration engendered construction of the hybrid domain at first with an irregular set of crossbred theoretical models. Eventually, step by step, on revealing and gradual eliminating the contradictions between the models involved, the hybrid set was put into order with a help of the equivalence principle. A hierarchy of theoretical models starting from the crossbreeds and up to usual hybrids was moulded. With the metric tensor at the top of the edifice Einstein was able to comprise both the knowledge on gravitation and inertia represented by classical mechanics and the knowledge on the structure

of space and time embodied by special relativity.

Einstein's unification design could be successfully implemented since his program masterly embraced the valuable ideas of the Nordström research program, as well as some enticing presuppositions of the program of Max Abraham. Only as a result of reconciling and amalgamating the 'physical' and 'mathematical' approaches, embodied in Abraham, Einstein and Nordström's crossbred theoretical models, Einstein was able to explain successfully the anomalous motion of Mercury.

The theories of Nordström and Abraham contrived before November 25, 1915, were not merely the scaffolds to construct the GTR basic model. They constitute the necessary part of the whole GTR edifice necessary for its common use. Notwithstanding Einstein's stupendous impact, the contributions of Nordström, Abraham, Poincare, Hilbert, Besso, Fokker and others should be taken into account.

8. To check the efficiency of the epistemic model, a peculiar process of unification of theories of electromagnetic and weak interactions that took place in the second half of the XX century is scrutinized. It is manifested that the epistemological specificity of the episode in the history of science is primarily expressed in the fact that the unification was carried out as a part of the substantially *synthetic* program that aimed to put up a sequence of intricate theories of electroweak interactions as consecutive approximations to the Yang-Mills gauge theory. The milestones of the implementation of the synthetic gauge program that culminated in the advancement of the so-called 'Glashow-Weinberg-Salam model', the dogged proof of its renormalizability and empirical adequacy are examined. A renowned tenet according to which the electroweak theory itself does not provide a full account of how the fields are mixed by furnishing direct values for the Weinberg angle, i.e. that the electroweak theory does not determine in any strict sense that the fields *must* be mixed is vehemently disputed. However, it is contended that in many important respects **the hybrid nature of this model engenders its unfinished nature.**

9. A fresh example of GTR and QFT encounter is scrutinized. It is contended that the substantive progress in their reconciliation was attained quite recently, and, at least at the first stages, it surprisingly

had nothing to do with superstrings, supergravity, and so on. Under the epistemic model, the real progress commenced with the hybrid model construction in the brand new crossbred realm called the 'Quantum Field Theory in Curved Spaces'. The crossbred realm embraces at least two important domains – black hole physics and inflationary cosmology. The titbits of the former are Hawking's path-breaking 1975 discovery of particle creation by black hole and Unruh's startling effect. The pivotal crossbreed here is the whimsical QFT vacuum experimentally approved in the 'Casimir effect'. When it was intruded to the GTR derivative system (the black hole physics), it did acquire a new and rather queer property – 'to produce infinitely great gravitational field'. **The introduction of the QFT vacuum restructured the relations between the GTR objects significantly, and the process of the global theory construction inevitably began.** The consequent inflation scenarios, the accelerating Universe queer models, and subtle superunification theories constitute the indispensable stages of the global theory construction.

Nevertheless, despite more than 30 years of stubborn research, the superstring theory remains not only empirically unconfirmed but also theoretically unfinished. The long list of the numerous superstring difficulties asserts that attainment of the consequent, verified Theory of Everything is apparently out of question.

REFERENCES

Abraham, Max (1912a)"Zur Theorie der Gravitation '. *Physikalische Zeitschrift"* 13 :1-4.

Abraham, Max (1912b)"Das Elementargesetz der Gravitation". *Physikalische Zeitschrift* 13:4-5.

Abraham, Max (1912c)"Der Freie Fall". *Physikalische Zeitschrift* 13: 310-311.

Abraham, Max (1912d)"Nochmals Relativitat und Gravitation. Bemerkung zu A. Einsteins Erwiderung". *Annalen der Physik* 38 : 1056-58.

Abraham, Max(1914)"Die neue Mechanik". *Scientia* 15 : 8-27.

Abraham, Max (1915)"Neuere Gravitationstheories". *Jahrbuch der Radioaktivität und Elektronik,* 470-520.

Alcoff, Linda M. (2001) "The Case for Coherence". In *The Nature of Truth. Classic and Contemporary Perspectives,* edited by Michael P. Lynch, 159-182. Cambridge: The MIT Press.

Bardeen J.M., B. Carter, and S.W. Hawking (1973) "The Four Laws of Black Hole Mechanics". *Communications in Mathematical Physics* 31:161-170.

Beller, Mara (2000) "Kant's Impact on Einstein's Thought". In *Einstein: The Formative Years, 1879-1909,* edited by Don Howard and John Stachel, 83-106. Boston: Birkhauser.

Birrell, N.D., and P.C.W. Davies (1982) *Quantum Fields in Curved Space.* Cambridge: Cambridge University Press.

Brown, Harvey R., and Rom Harre (1990) "Introduction". In *Philosophical Foundations of Quantum Field Theory,* edited by Harvey Brown and Rom Harre, 1-6. Oxford: Clarendon Press.

Blanschard, Brand (2001) "Coherence as the Nature of Truth". In *The Nature of Truth. Classic and Contemporary Perspectives*, edited by Michael P. Lynch, 103-122.Cambridge: The MIT Press.

Bohr, Niels (1913) "On the Constitution of Atoms and Molecules". *Philosophical Magazine* 26: 1-25; 476-502; 857-875.

Boyer, Timothy (1975) "Random electrodynamics: The theory of classical electrodynamics with classical electromagnetic zero-point radiation". *The Physical Review* 11, no 4:790-808.

Boyer, Timothy (1984) "Derivation of the black-body radiation spectrum from the equivalence principle in classical physics with classical electromagnetic zero-point radiation". *The Physical Review* D 29, no 6:1096-1097.

Brush, Stephen G. (1967) "Foundations of Statistical Mechanics, 1845-1915". *Archive for History of Exact Sciences* 4: 145-183.

Buchwald, Jed (1998) The *creation of scientific effects: Heinrich Hertz and electric waves*. Chicago: The University of Chicago Press.

Bucklaev, V.G. (1957) "L. Boltzmann and Planck's quantum hypothesis". *Voprosi Istorii Yestestvoznaniya i Techniki* (The Problems of History of Natural Sciences and Techniques) 4: 167-168 (in Russian).

Cao, Tian Yu. (1997) *Conceptual Developments of 20th century field theories*. Cambridge: Cambridge University Press.

Cartwright, Nancy (1985) *How the Laws of Physics Lie*. Cambridge: Cambridge University Press.

Chyba, C.F. (1985) "Kaluza-Klein unified theory and apparent 4-dimensional space-time". *American Journal of Physics* 53(9): 863-872.

Darrigol, Olivier (2001) *Electrodynamics from Ampere to Einstein*. Oxford: Oxford University Press.

Dawid, Richard (2013) *String Theory and the Scientific Method*. Cambridge: Cambridge University Press.

De Broglie, Louis (1962) "Mysterious constant h - Max Planck's celebrated discovery". In *Following the Paths of Science*, edited by Yuri V. Sachkov, 139-146. Moscow: Izdatelstvo Inostrannoy Literaturi (Foreign Literature Publishing House, in Russian).

De Witt, B.S. (1975) "Quantum-Field Theory in Curved Space-times". *Physics Reports* C.19: 297-357.

De Witt, B.S. (1980): "Quantum Gravity: the New Synthesis".In *Quantum Gravity 2, 2nd Oxford Symposium on Quantum Gravity*, edited by C.J. Isham, Roger Penrose and Dennis William Sciama, 680-745. Oxford : Oxford University Press.

Dugas, Rene (1959): *La theorie physique au sens de Boltzmann et ses prolongements modernes.* Paris: Editions de Griffon.

Duhem, Pierre ([1906], 1954) *The Aim and Structure of Physical Theory.* Princeton, New Jersey: Princeton University Press.

Einstein, A. (1902)"Über die thermodynamische Theorie der Potentialdifferenz zwischen Metallen und vollständig dissoziierten Lösungen ihrer Salze, und eine elektrische Methode zur Erforschung der Molekularkräfte". *Annalen der Physik* 8: 798-814. Translated by Anna Beck in: The Collected papers of Albert Einstein. Vol. 2. The Swiss years: writings, 1900 – 1909. Princeton, New Jersey: Princeton University Press, 1989, 12-29.

Einstein, A. (1903)"Eine Theorie der Grundlagen der Thermodynamik". *Annalen der Physik* 11: 170-187. Translated by Anna Beck in: The Collected Papers of Albert Einstein. Vol. 2. The Swiss years: writings, 1900 – 1909. Princeton, New Jersey: Princeton University Press, 1989, 48-67.

Einstein, A. (1904)"Zur allgemeinen molekularen Theorie der Warme". *Annalen der Physik* 14 : 354-362. Translated by Anna Beck in: The Collected Papers of Albert Einstein. Vol. 2. The Swiss years: writings, 1900 – 1909. Princeton, New Jersey: Princeton University Press, 1989, 68-77.

Einstein, A. (1905a)"Über einen die Erzeugung und Verwandlung des Lichtes betreffenden heuristischen Gesichtspunk". *Annalen der Physik* 17:132-48. Translated by Anna Beck in: The Collected Papers of Albert Einstein. Vol. 2. The Swiss years: writings, 1900 – 1909. Princeton, New Jersey: Princeton University Press, 1989, 86-103.

Einstein, A. (1905b)"Eine neue Bestimmung der Moleküldimensionen". University of Zürich Dissertation. Translated by Anna Beck in: The Collected Papers of Albert Einstein. Vol. 2. The Swiss years: writings, 1900 – 1909. Princeton, New Jersey: Princeton University Press, 1989, 104-122.

Einstein, A. (1905c)"Über die von der molekularkinetischen Theo-

240

rie der Wärme geforderte Bewegung von in ruhenden Flüssigkeiten suspendierten Teilchen". *Annalen der Physik* 17: 549-560.Translated by Anna Beck in: The Collected Papers of Albert Einstein. Vol. 2. The Swiss years: writings, 1900 – 1909. Princeton, New Jersey: Princeton University Press, 1989, 123-134.

Einstein, A. (1905d) "Zur Elektrodynamik bewegter Körper" *Annalen der Physik* 17: 891-921.English translation in: *The Principle of Relativity*, Dover, New York, 1923. See also the translation of Anna Beck in: *The Collected Papers of Albert Einstein.* Vol. 2. The Swiss years: writings, 1900 – 1909. Princeton, New Jersey: Princeton University Press, 1989, 140 - 171.

Einstein, A. (1905e)"Ist die Trägheit eines Körpers von seinem Energienhalt abhängig? *"Annalen der Physik* 18:639-641. Translated by Anna Beck in: The Collected Papers of Albert   Einstein. Vol. 2. The Swiss years: writings, 1900 – 1909. Princeton, New Jersey: Princeton University Press, 1989, 171 - 174.

Einstein, A. (1907)"Über das Relativitätsprinzip und die aus demselben gezogenen Folgerungen". *Jahrbuch der Radioaktivität und Elektronik*: 4, 411- 462. Translated by Anna Beck in: The Collected Papers of Albert Einstein. Vol. 2. The Swiss years: writings, 1900 – 1909. Princeton, New Jersey :Princeton University Press, 1989, 252-311.

Einstein, A. (1909)"Über die Entwicklung unserer Anschauungen über das Wesen und die Konstitution der Strahlung". *Physikalische Zeitschrift* 10:817-25. Translated by Anna Beck in: The Collected Papers of Albert Einstein. Vol. 2. The Swiss years: writings, 1900 – 1909. Princeton, New Jersey: Princeton University Press, 1989, 379 - 394.

Einstein, A. (1910) "Principé de relativité et ses consequencé dans la physique modern". *Archives des Sciences Physique et Naturelles* 29 :125-44.

Einstein, Albert (1911 a)"Zur Theorie des statischen Gravitationsfeldes". *Annalen der Physik* 38 :443-458.

Einstein, Albert. (1912b)"Relativität und Gravitation. Erwiderung auf eine Bemerkung von M. Abraham", *Annalen der Physik* 38 :1059-1064.

Einstein, Albert, and Marcel Grossmann(1913a)"Entwurf einer verallgemeinerten Relativitätstheorie und Theorie der Gravitation". *Zeitschrift für Mathematik und Physik* 61 : 225-261.

Einstein, Albert(1913b)"Physikalische Grundlagen einer Gravitationtheorie". *Naturforchende Gesellschaft*, Zürich, Vierteljahrsschrift 58 :284-290.

Einstein, Albert (1913c)"Zum gegenwärtigen Stande des Gravitationsproblems". *Physikalische Zeitschrift* : 1249-1262. Reprinted in: Jürgen Renn,ed. *The Genesis of General Relativity. Vol.3. Gravitation in the Twilight of Classical Physics: Between Mechanics, Field Theory and Astronomy.* Dordrecht: Springer, 2007.

Einstein, Albert, and Adrian Fokker (1914)"Die Nordströmsche Gravitationtheorie vom Standpunkt des absoluten Differentialkalkulis". *Annalen der Physik* 44, pp.321-28.

Einstein, Albert (1915)"Die Feldgleichungen der Gravitation". *Preußische Akademie der Wissenschaften, Sitzungsberichte* 48 : 844-847.

Einstein, Albert (1916)"Näherungsweise Integration der Feldgleichungen der Gravitation". *Preußische Akademie der. Wissenschaften, Sitzungsberichte* 1 :688-696.

Einstein, A. (1918)"Motive des Forschens". In *Zu Max Planks Sechszigstem Geburtstag.* Ansprachen, gehalten am 26 April 1918 in der Deutschen Physikalischen Gesellschaft, 29-32, Karlsruhe: Muller.

Einstein, A. ([1933],1954)"On the Method of Theoretical Physics". Reprinted in: *Ideas and Opinions,* 277-283. New York: Crown Publishing Inc.

Einstein, Albert ([1936], 1954) "Physics and Reality". *The Journal of the Franklin Institute* 221, no 3 (March). Reprinted in: Albert Einstein. *Ideas and Opinions.* Translated by Sonja Bargmann. N.Y.: Crown Publishers, 1960, 290-323.

Einstein, A. ([1940], 1954) "The Fundaments of Theoretical Physics", In *Ideas and Opinions*, 323-330, New York: Crown Publishing Inc.

Einstein, A. ([1944], 1954)"Remarks on Bertrand Russell's Theory of Knowledge". In: *The Philosophy of Bertrand Russell,* edited by Paul Arthur Schilpp, 278-292. Evanston, Illinois: The Library of Living Philosophers,. Reprinted in *Ideas and Opinions*, 18-24, New York:

Crown Publishers Incorp.

Einstein, A. (1949a) "Autobiographical Notes". In *Albert Einstein: Philosopher -Scientist*, edited by P.A. Schilpp, vols 1-2, 1-14, Evanston: IL.

Einstein, A. (1949b)"Remarks Concerning the Essays Brought together in this Co-operative Volume". In *Albert Einstein: Philosopher - Scientist*, edited by P.A. Schilpp, vols 1-2, 665-688, Evanston: IL.

Einstein, A. (1954)"Relativity and the Problem of Space". In *Ideas and Opinions*, 360-370, New York: Crown Publishing Inc.

Einstein, A. (1987) The Collected Papers of Albert Einstein. Vol. 1.The Early Years, 1879-1902. Edited by John Stachel. Princeton: Princeton University Press.

Einstein, A. (1989) The Collected papers of Albert Einstein. Vol. 2. The Swiss years: writings, 1900 – 1909. Princeton, New Jersey: Princeton University Press,.

Albert Einstein / Mileva Marić (1992) *The Love Letters*. Edited and with an introduction by Jürgen Renn and Robert Schulmann. Translated by Syawn Smith. Princeton, NJ: Princeton University Press.

Einstein, A. (1993) The Collected papers of Albert Einstein. Vol. 5. The Swiss years: writings, 1901 – 1914. Princeton, New Jersey: Princeton University Press.

Fermi, E. (1934)"Versuch einer Theorie der β- Strahlen". *Zeitschrift fur Physik* 88 :161-171. Translated by Wilson F.L. (1968) "Fermi's Theory of Beta Decay". *American Journal of Physics* 36, no 12: 1150-1160.

Feuer, Louis S. (2017) *Einstein and the Generations of Science.* New York: Routledge.

Feyerabend, Paul (1975) *Against Method.* London: New Left Books.

Feynman, R.P., and M. Gell - Mann (1958) "Theory of the Fermi Interaction". *The Physical Review* 109:193-198.

Foucault, Michel (2001) "Truth and Power". In *The Nature of Truth. Classic and Contemporary Perspectives,* edited by Michael Lynch, 317-320. Cambridge: The MIT Press.

Frank, Ph. (1949)"Einstein, Mach and Logical Positivism". In Albert *Einstein: Philosopher-Scientist,* edited by P.A. Schilpp, vols

1-2, 271-286. Evanston: IL.

Fulling, S.A. (1973)"Nonuniqueness of Canonical Field Quantization in Riemannian Space-Time". *The Physical Review* 7: 2850-2861.

Galison, Peter (1987) *How Experiments End.* Chicago: The University of Chicago Press.

Galison, P., and D. Stump (1996) The *Disunity of Science.* Stanford: Stanford University Press.

Glashow, S.L. (1961)"Partial Symmetries of Weak Interactions". *Nuclear Physics* 22: 579-588.

Goldhammer, D.A. (1911)"New Ideas in Modern Physics". *Fizicheskoye Obozreniye* (The Review of Physics) 12, no 1-2:1-35, 76-155 (in Russian).

Gould, A. (1987) "Classical derivation of black-hole entropy". *The Physical Review* D35, no 2: 449-454.

Green, Brian (2000) *The Elegant Universe. Superstrings, Hidden Dimensions and the Quest for Ultimate Theory.* New York: Vintage Books.

Greiner W., and B. Muller (2009) Gauge *Theory of Weak Interactions.* Heidelberg – Dordrecht: Springer.

Hacking, Ian (1985) Representing *and Intervening.* Cambridge: Cambridge University Press.

Hawking, S.W. (1975)"Particle Creation by Black Holes". *Communications in Mathematical Physics* 43: 199-220.

Heidegger, Martin (1971) *Was heist Denken?* Tübingen: Max Niemeyer Verlag.

Heidegger, Martin (1972)"Die Zeit des Weltbildes". In *Holzwege.* Frankfurt: V.Klostermann. Heidegger, Martin (1978) "Die Frage nach der Technik". *Humanitas* 14: 17-21.

Heidegger, Martin (1995) *The Fundamental Concepts of Metaphysics:World, Finitude, Solitude.* Bloomington: Indiana University Press.

Heidegger, Martin (1996) *Being and Time.* Albany: State University of New York. Translated by Joan Stambough from Heidegger's *Zein und Zeit.* Max Niemeyer, 1927.

Heidegger, Martin (2001) "On the Essence of Truth". In *The Nature of Truth,* edited by Michael Lynch, 295-316. Cambridge: The MIT

Press.

Hentschel, Klaus (2018) *Photons: the History and Mental Models of Light Quanta.* Springer.

Hertz, H. (1884)"Über die Beziehungen zwischen den Maxwell'schen elektrodynamischen Grundgleichungen und den Grundgleichungen der gegnerischen Elektrodynamik". *Annalen der Physik und Chemie* 23: 84-103.

Hertz, Heinrich (1893) *Electric Waves.* London: Macmillan.

Hertz, Heinrich (1899) The *Principles of Mechanics presented in a new form.* Translated by D.E. Jones. London: Macmillan.

Higgs, P.W. (1964) "Broken Symmetries, Massless Particles and Gauge Fields". *Physics Letters* 12: 132-133.

Higgs, P.W. (1966) "Spontaneous Symmetry Breaking Without Massless Bosons". *The Physical Review* 145: 1156-1163.

Holton, Gerald (1968)"Mach, Einstein and the Search for Reality". *Daedalus* 97: 636-673.

Hossenfelder, Sabine (2020) *Lost in Math: How Beauty Leads Physics Astray.* New York: Basic Books.

Howard, Don (1994)"Einstein, Kant and the Origins of Logical Positivism". In *Language, Logic and the Structure of Scientific Theories: The Carnap -Reichenbach Centennial,* edited by Wesley Salmon and Gedeon Wolters, 45-105. Pittsburgh: University of Pittsburgh Press.

Howard, Don and John Stachel (eds.) (2000) *Einstein. The Formative Years, 1879-1909.* Boston: Birkhauser.

Howard, Don (2014)"Einstein and the Development of Twentieth-Century Philosophy of Science". In: *Cambridge Company to Einstein,* edited by Michel Janssen and Christoph Lehner, 354-376. Cambridge: Cambridge University Press.

Hoyer, Ulrich (1980) "Von Boltzmann zu Planck". *Archive for History of Exact Sciences* 23, no 1 : 47-86.

Jammer, Max (1966) *The Conceptual Development of Quantum Mechanics.* New York: Mc Graw-Hill.

Jeans, James (1905)"On the Partition of Energy between Matter and Aether".*Philosophical Magazine* 10: 91-98.

Kangro, Hans (1970) *Vorgeschichte des Planck'schen Strahlungsgesetzes.* Wiesbaden: Steiner. Translated as History *of Planck's Ra-*

diation Law. London: Taylor and Francis, 1976.

Kant, Immanuel ([1781], 1998) The *Critique of Pure Reason*. Translated by Paul Guyer and Allen W. Wood. Cambridge: Cambridge University Press.

Kant, I. ([1783], 2002) *Prolegomena to any future metaphysics that will be able to come forward as science*. Repr. Translated by Gary Hatfield. In: *Immanuel Kant. Theoretical Philosophy after 1781*. Cambridge: Cambridge University Press.

Kelvin, Lord (1901)"19-th Century Clouds over the Dynamical Theory of Heat and Light". Philosophical Magazine 2:1-39.

Klein. Martin J. (1962) "Max Planck and the beginning of the quantum theory". *Archive for History of Exact Science* 1, no 5: 459-479.

Klein, Martin J. (1966)"Thermodynamics and quanta in Planck's work". *Physics Today* 9, no 11: 23-28.

Kobzarev, Igor Yu. (1979)"A. Einstein, M. Planck and atomic theory". *Priroda* (Nature), no 3: 8-26 (in Russian).

Kragh, Helge S. and James M. Overduin (2014) *The Weight of the Vacuum: A Scientific History of Dark Energy*. Heidelberg: Springer.

Kuhn, T.S. (1958) "Newton's Optical Papers". In *Isaac Newton's Papers and Letters on Natural Philosophy*, edited by I. Bernard Cohen, 27-45. Cambridge, Massachusetts: Harvard University Press.

Kuhn, T.S. (1963) "The Function of Dogma in Scientific Research". In *Scientific Change*, edited by A.C. Crombie, 347-359.Cambridge: Massachusetts: Harvard University Press.

Kuhn, T.S. (1973)"Objectivity, Value Judgement and Theory Choice". In *The Essential Tension*, 320 - 339. Chicago: University of Chicago Press.

Kuhn, T.S. (1978) *Blackbody Theory and Quantum Discontinuity, 1894-1912*. Oxford and New York: Oxford University Press.

Kuhn, Thomas S. (1980) "The Halt and the Blind: Philosophy and History of Science". *The British Journal for the Philosophy of Science* 31: 181-192.

Kuhn, Thomas S. ([1961], 1996). *The Structure of Scientific Revolutions*. 3rd Edition. Chicago: The University of Chicago Press.

Kuhn, Thomas S. (2000) *The Road Since Structure.* Chicago: The University of Chicago Press.

Lakatos, Imre (1970)"Falsification and the Methodology of Scientific Research Programmes". In *Criticism and the Growth of Knowledge,* edited by I. Lakatos and A. Musgrave, 91-195. Cambridge: Cambridge University Press.

Lakatos, Imre (1971)"History of Science and its Rational Reconstructions". In Boston *Studies in the Philosophy of Science, vol.8,* edited by R.C. Buck and R.S. Cohen. Reprinted in *Method and Appraisal in the Physical Sciences,* edited by C. Howson, 1976. Cambridge: Cambridge University Press.

Landau, Lev (1958)"The Quantum Theory from Max Planck up to now". In *Max Planck.1858-1958,* 93-98. Moscow: USSR Academy of Science Publishing House (in Russian).

Landau, Lev D. and Yevgeny M. Lifshitz (1973) *Theory of Fields.* London: Pergamon Press.

Larmor, J. (1909)"On the Statistical and Thermodynamical Relations of Radiant Energy". *Proceedings of the Royal Society* A560: 82-95.

Lenzen, V.F. (1949) "Einstein's Theory of Knowledge". In: *Albert Einstein: Philosopher-Scientist,* edited by P.A. Schilpp, 357-384. Volumes 1-2, Evanston, IL.

Lorentz, H.A. (1909) *The theory of electrons and its applications to the phenomena of light and radiant heat.* New York: G.F. Stechart & Co.

Laudan, Larry (1006) "A Confutation of Convergent Realism". In *The Philosophy of Science,* edited by David Papineau, 107-138. Oxford: Oxford University Press.

Mach, Ernst ([1893], 1999) The *Science of Mechanics: A Critical and Historical Account of its Developments.*Translated by T. J. Mc Cormmach. La Salle: Open Court.

Mach, E. ([1897], 1984) The *Analysis of Sensations and the Relation of the Physical to the Psychical.* Translated by C.M. Williams. La Salle: Open Court.

Maldacena J., A. Strominger and E. Witten (1997) "Black-Hole entropy in M-theory". *Journal of High Energy Physics* 2: 209-254.

247

Maxwell, J.C. ([1873a], 1952)"On Action at a Distance". *Proceedings of the Royal Institute GB* VII: 48-49. Reprinted in: *The Scientific Papers of James Clerk Maxwell*, 1890, vol.1, Cambridge: The University Press and N.Y: Dover, 1952: 315-20.

Maxwell, James ([1873b], 1954) A *Treatise on Electricity and Magnetism*. 1 vols, third ed. New York: Dover.

Maxwell, James ([1877], 1952) "Hermann Ludwig Ferdinand Helmholtz", *Nature* XV: 389-391. Reprinted in: *The Scientific Papers of James Clerk Maxwell*, 1890, vol.2, Cambridge: The University Press and N.Y: Dover, 1952, 592-659.

Miller, A. (1981) Albert *Einstein's Special Theory of Relativity*. London: Addison-Wesley. Norton, John (1993)"The Determination of Theory by Evidence: The Case for Quantum Discontinuity, 1900-1915". *Synthese* 97: 1-31.

Morrison, Margaret (2007) *Unifying Scientific Theories. Physical Concepts and Mathematical Structures*. Cambridge: Cambridge University Press.

Nambu, Y., and G. Jona - Lasilio (1961) "Dynamical Model of Elementary Particles Based on an Analogy with Superconductivity". *The Physical Review* 122: 345-358; 124-246.

Nordström, Gunnar (1912) "Relativitätsprinzip und Gravitation". *Physikalische Zeitschrift* 13: 1126-1129. Translated in Jürgen Renn, ed., The Genesis of General Relativity. Vol.3.Gravitation in the Twilight of Classical Physics: Between Mechanics, Field Theory and Astronomy. Dordrecht : Springer, 2007.

Nordström, Gunnar(1913a)"Trage und Schwere Masse in der Relativitätsmechanik". *Annalen der Physik* 40: 856-878.

Nordström, Gunnar (1913b)"Zur Theorie der Gravitation vom Standpunkt des Relativitätsprinzip". *Annalen der Physik* 42:533-534.

Norton, J. D. (2010) "How Hume and Mach Helped Einstein Find Special Relativity". In Discourse *on a New Method. Reinvigorating the Marriage of History and Philosophy of Science*, edited by M. Dickson and M. Domski, 359-386. Chicago and La Salle, IL: Open Court.

Nugayev, Rinat (1985a) "A Study of Theory Unification". *The British Journal for the Philosophy of Science* 36:159-79.

Nugayev, R.M. (1985b) "The History of Quantum Theory as a

Decisive Argument Favoring Einstein over Lorentz". *Philosophy of Science* 52:44-63.

Nugayev, Rinat (1996) "Why did new physics force out the old?" *International Studies in the Philosophy of Science* 10, no 2:127-140.

Nugayev, Rinat M. (1999) *Reconstruction of Mature Theory Change: A Theory –Change Model.* Peter Lang: Frankfurt am Main.

Nugayev, Rinat (2013) "The Ptolemy-Copernicus Transition: Intertheoretical Context". *Almagest* 4(1): 96-119.

Nugayev, Rinat (2015) "Communicative Rationality of the Maxwellian Revolution". *Foundations of Science* 20, issue 4: 447-478.

Nugayev, Rinat and Tanzilia Burganova(2016) "Epistemologia di Heidegger come racconto di source Kuhn crescita della conoszenza scientifica". *Italian Science Review* 1(34) :156-167.

Nugayev, Rinat M. (2018a) *Einstein's Revolution: A Study in Theory Unification.* Bentham Science Publishers: Sharjah, UAE

Nugayev, Rinat M. (2018b) "The Genesis of General Relativity: An Inter-Theoretical Context". *Acta Baltica Historiae et Philosophiae Scientiarum* 6, no 1:97-129.

Okun, L.B. (1987) *Leptons and Quarks.* Amsterdam: Elsevier Science Publishers.

Pais, A. (1979) "Einstein and the Quantum Theory". *Reviews of Modern Physics* 51: 863-914. Pais, A. (1982) *Subtle is the Lord. The Science and Life of Albert Einstein.* Oxford: Oxford University Press.

Papineau, David (1996) "Introduction". In *The Philosophy of Science,* edited by David Papineau, 3-22. Oxford: Oxford University Press.

Perlmutter, S. et al. (1999) "Measurements of ω and λ from 41 high-redshift supernovae'. *The Astrophysical Journal* 517:565-586.

Pickering, Andrew (1985) *Constructing Quarks. A Sociological History of Particle Physics.* Chicago: The University of Chicago Press.

Pierce, C.S. (2001) "How to Make Our Ideas Clear". In *The Nature of Truth. Classic and Contemporary Perspectives,* edited by M. Lynch, 193-210. Cambridge: The MIT Press.

Planck, Max (1897a)"Uber irreversible Strahlungsvorgange. Erste Mitteilung". *Berichte der Deutsche Chemischen Gesellchaft* 57-68.

Planck, Max (1897b) "Uber irreversible Strahlungsvorgange. Zweite Mitteilung". *Berichte der Deutsche Chemischen Gesellschaft*,715-717.

Planck, Max (1897c) "Uber irreversible Strahlungsvorgange. Dritte Mitteilung". *Berichte der Deutsche Chemischen Gesellschaft* 1122-1145.

Planck, Max (1898) "Uber irreversible Strahlungsvorgange. Vierte Mitteilung". *Berichte der Deutsche Chemischen Gesellchaft* 57-68.

Planck, Max (1899) "Uber irreversible Strahlungsvorgange. Fünfte Mitteilung (Schluss)". *Berichte der Deutsche Chemischen Gesellchaft* 440-480.

Planck, Max (1900a)"Uber irreversible Strahlungsvorgange". *Annalen der Physik* 1 : 69-122. Planck, Max (1900b)"Zur Theorie des Gesetzes der Energieverteilung im Normalspectrum". *Verh. D. D. Phys. Ges.*2: 237-245. Translated in Ter Haar, 1967.

Planck, Max (1906) *Vorlesungen über die Theorie der Wärmestrahlung.* First edition. Leipzig: J.A. Barth.

Planck, Max (1910) *Acht Vorlesungen über theoretische Physik.* Leipzig: Hirzel. Originally delivered at Columbia University, 1909, and translated as "Eight Lectures on Theoretical Physics" by A.P. Wills. New York: Columbia University Press, 1915.

Planck, Max (1911) "Eine Neue Strahlungshypothes". *Verh. Dtsch. Phys. Ges.* 13:138-148.

Planck, Max (1931) *Max Planck and Robert William Wood. Brief 1931.* Sources for History of Quantum Physics. Mf. 66,7x.

Planck, Max (1960) *A Survey of Physical Theory.* Translated by R. Jones and D.H. Williams. NY: Dover Publications.

Podgoretzky, M., and J. Smorodinsky (1980)"On the Axiomatic Structure of Physical Theories". In The *Physical Theory*, edited by Yu. V. Sachkov, 53-61.Moscow: Nauka (in Russian).

Poincaré, H. ([1902], 1905) Science *and Hypothesis.* Repr.London and Newcastle-on-Tyne: The Walter Scott Publishing Co.

Poincaré, Henri (1906) Science *et Hypothese.* Paris: Flammarion.

Popper, Karl R. (1963) *Conjectures and Refutations: The Growth of Scientific Knowledge.* London: Routledge & K. Paul.

Psillos, Stasis (1995) "Is Structural Realism the Best of Both Worlds?" *Dialectica* 49:5-46.

250

Putnam, Hilary (2001) "Two Philosophical Perspectives". In *The Nature of Truth. Classic and Contemporary Perspectives*, edited by M. Lynch, 251-258. Cambridge, Massachusetts: The MIT Press.

Quine, W.V. (2000) "Two Dogmas of Empiricism". In *Contemporary Analytic and Linguistic Philosophies*, edited by E.D. Klemke, 331-349. New York: Prometheus Books.

Renn, Jürgen and Sauer, Tilman (2007) "Pathways out of Classical Physics: Einstein's Double Strategy in his Search for the Gravitational Field Equations". In *The Genesis of General* Relativity, edited by Jürgen Renn, 113-312. Vol. 1. Dordrecht: Springer.

Renn, Jürgen (2007a) "Classical Physics in Disarray. The Emergence of the Riddle of Gravitation", In *The Genesis of General Relativity*, edited by Jürgen Renn, 21-84. Vols. 1-2, Dordrecht: Springer.

Renn, Jürgen. (2007b) "The summit almost scaled: Max Abraham as a pioneer of a relativistic theories of gravitation", In *The Genesis of General Relativity*, edited by Jürgen Renn, 305-330. Vol.3.Gravitation in the Twilight of Classical Physics: Between Mechanics, Field Theory and Astronomy. Dordrecht: Springer.

Rovelli, Carlo (2008) Quantum *Gravity*. Cambridge: Cambridge University Press.

Rynasiewicz, R. (2000) "The Construction of the Special Theory: Some Queries and Considerations". In *Einstein. The Formative Years, 1879-1909*, edited by Don Howard and John Stachel, 159-201, Boston: Birkhauser.

Rynasiewicz, R. (2005) "The optics and electrodynamics of 'On the Electrodynamics of Moving Bodies'". In *Einstein's Annalen Papers. The Complete Collection 1901-1922*, edited by Jürgen Renn. Germany: Wiley-VCH Verlag GmbH&Co.

Ryckman, T. (2005) *The Reign of Relativity: Philosophy in Physics 1915-1925*. Oxford: Oxford University Press.

Salam, A., and J.C. Ward (1964) "Electromagnetic and Weak Interactions". *Physics Letters* 13:168-171.

Salam, A. (1984a) "Einstein's Last Dream: The Space-Time Unification of Fundamental Forces". In: *Abdus Salam. Ideals and Realities*, 299-309. Singapore: World Scientific.

Salam, A. (1984b) "The Nature of the "Ultimate" Explanation in

Physics". In *Abdus Salam. Ideals and Realities*, 310-319. Singapore: World Scientific,

Salam, Abdus (1984c) "Gauge Unification of Fundamental Forces". In: *Abdus Salam. Ideals and Realities*, 320-369. Singapore: World Scientific.

Sauer, Tilman (2015) "Marcel Grossmann and his Contribution to the General Theory of Relativity". In Proceedings of the 13th Marcel Grossmann Meeting, Singapore: World Scientific, 456-503.

Schwinger, J. A. (1957) "Theory of Fundamental Interactions". *Annals of Physics* 2: 407-434.

Seelig, Carl(1957) *Helle Zeiten – Dunkle Zeiten : In memoriam Albert Einstein*. Zürich /Stuttgart/ Wien: Europa Verlag.

Seelig, Carl (1960) *Albert Einstein. Leben und Werk Eines Genies Unserer Zeit*. Europa Verlag.

Slavov, M. (2016) "Empiricism and Relationism Intertwined: Hume and Einstein's STR".*Theoria* 31/2: 247-263.

Smolin, Lee (2006) *The Trouble with Physics*. Boston-New York: Houghton-Mifflin.

Speziali, P., editor (1972) *Albert Einstein-Michele Besso: Correspondence*. Paris: Hermann.

Stachel, J. (2000)" Introduction to Einstein: The Formative Years". In *Einstein. The Formative Years, 1879-1909*, edited by Don Howard and John Stachel, 1-21. Boston: Birkhauser.

Stachel, J. (2008) "Albert Einstein". In *The New Dictionary of Scientific Biography*, edited by G. Gale, vol.2.

Stepin, Vyacheslav (2005) *Theoretical Knowledge*. Dordrecht-Reidel: Springer.

Suppe, Frederick [1974], (1977) The Structure of Scientific Theories. The University of Illinois Press, 1977.

Ter Haar, Dirk (1967) *The Old Quantum Theory*. Oxford: Pergamon Press.

t'Hooft, G. (1980) "Gauge Theories of the Forces between Elementary Particles". *Scientific American* 242(6): 90-119.

Tisza, L. (1963)"The Conceptual Structure of Physics". *Reviews of Modern Physics* 35:151-185.

Tolman, R. (1912) "Some Emission Theories of Light". *Physical Review* 35:136-143.

Van Dongen, J. (2010) Einstein's *Unification*. Cambridge: Cambridge University Press.

Unruh, W.G. (1976)"Notes on black-hole evaporation". *The Physical Review* D14, no 4.

Unruh, W.G. (1980)"Comment on Hawking's paper". In: *Some Strangeness in the Proportion*, edited by Harry Wolf, 153-155. N.Y.: Addison-Wesley.

Van Dongen, Jeroen (2010) Einstein's *Unification*. Cambridge: Cambridge University Press.

Vihalemm, Rein and Peter Müürsepp (2007) "Philosophy of Science in Estonia", *Journal for General Philosophy of Science* 38(1):167-191.

Vizgin, V.P. (1981) *Relativistic theory of gravity: genesis and development. 1900-1915.*Moscow: Nauka (in Russian).

Vladimirov, Yu. V. (1989) *Space-time: explicit and implicit dimensions.* Moscow: Nauka (in Russian).

Walker, Ralph C.S. (2001) "The Coherence Theory". In *The Nature of Truth. Classic and Contemporary Perspectives*, edited by M. Lynch, 123-158. Cambridge: The MIT Press.

Wartofsky, Marx W. (1979) *Models. Representation and the Scientific Understanding.* Springer: Dordrecht.

Weber, Max (1964) *Gesammelte Aufsätze zur Wissenschaftslehre.* Tübingen: Mohr.

Whewell, W. (1847) *The Philosophy of the Inductive Sciences, founded upon their history.* Vol. 2. London: John L. Parker.

Weinberg, Steven (1967) "A Model of Leptons'. *The Physical Review Letters* 19:1264-1266.

Weinstein, G. (2013) "The 1905 Relativity Paper and the "Light Quantum". *arXiv*.1307.2131 v1 [physics. hist-ph]

Wien, Wilhelm (1896) "On the Division of Energy in the Emission-Spectrum of a Black Body", translated in *Philosophical Magazine* 43, 1897, 214-220.

Woigt, Peter (2006) *Not Even Wrong: The Failure of String Theory and the Search for Unity in Physical Law.* N.Y.: Basic Books.

Worrall, John (1989) "Fresnel, Poisson and the white spot: the role of successful predictions in the acceptance of scientific theories". In *The Uses of Experiment*, 135-157. Cambridge: Cambridge University Press.

Yang, C.N., and Robert Mills (1954) "Conservation of Isotopic Spin and Isotopic Gauge Invariance". *The Physical Review*: 96-191.

Zel'dovich Ya.B. (1968) "The cosmological constant and the theory of elementary particles". *Soviet Physics Uspekhi* 11: 381-393.

Zel'dovich Ya. B., and I.D.Novikov (1973) *Stars and Relativity*. London: Pergamon Press.

www.ingramcontent.com/pod-product-compliance
Lightning Source LLC
Chambersburg PA
CBHW071723200326
41519CB00021BC/6548